Risk, Society and Policy

The Tolerability of Risk

Risk, Society and Policy Series
Series editor: Ragnar E. Löfstedt

The Tolerability of Risk

A New Framework for Risk Management

Edited by
Frédéric Bouder, David Slavin
and Ragnar E. Löfstedt

earthscan
from Routledge

First published by Earthscan in the UK and USA in 2007

2 Park Square, Milton Park, Abingdon, Oxfordshire OX14 4RN
52 Vanderbilt Avenue, New York, NY 10017

Routledge is an imprint of the Taylor & Francis Group, an informa business

First issued in paperback 2018

ISBN 978-1-84407-398-6 (hbk)
ISBN 978-1-84407-609-3 (pbk)

Typeset by FiSH Books, Enfield
Cover design by Yvonne Booth

A catalogue record for this book is available from the British Library

Library of Congress Cataloging-in-Publication Data

The tolerability of risk : a new framework for risk management / edited by Frédéric Bouder, David Slavin and Ragnar E. Löfstedt.
    p. cm.
  ISBN-13: 978-1-84407-398-6 (hardback)
  ISBN-10: 1-84407-398-X (hardback)
  1. Risk management. 2. Risk assessment. 3. Health risk assessment. 4. Safety regulations. I. Bouder, Frédéric. II. Slavin, David. III. Löfstedt, Ragnar.
  HD61.T65 2007
  338.5—dc23
                                                    2006100481

Contents

List of Figures and Tables

Figures

Tables

List of Acronyms and Abbreviations

ACOPs   approved codes of practice
ACSHH   Advisory Committee for Safety, Hygiene and Health Protection at Work
ACTS   Advisory Committee on Toxic Substances
ALARA   as low as reasonably achievable
ALARP   as low as reasonably practicable
BACT   best available control technology
BAT   best available technology
BATNEEC   best available techniques not entailing excessive cost
BSE   bovine spongiform encephalopathy
CBA   cost–benefit analysis
CBI   Confederation of British Industry
CEN   European Committee for Standardization (Comité Européen de Normalisation)
EMEA   European Medicines Agency
EMM   Enforcement Management Model
EP   exceedance probability
FDA   Food and Drug Administration
FSA   Food Standards Agency
GIS   geographic information systems
GMO   genetically modified organism
HD   Health Department
HSC   Health and Safety Commission
HSE   Health and Safety Executive
HSWA   Health and Safety at Work etc Act of 1974
IPCC   Intergovernmental Panel on Climate Change
IPCS   International Programme on Chemical Safety
IRGC   International Risk Governance Council
ISO   International Organization for Standardization
MEL   maximum exposure level
MHRA   Medicines and Healthcare products Regulatory Agency
MMR   measles, mumps and rubella
NOAEL   no adverse effect level
NUSAP   numeral, unit, spread, assessment, pedigree
OEL   occupational exposure standard

PRA	probabilistic risk assessment
PRIMA	pluralistic framework of integrated uncertainty management and risk analysis
QRA	quantified risk assessment
ToR	tolerability of risk
TUC	Trades Union Congress
WATCH	Working Group on the Assessment of Toxic Chemicals
WBGU	German Council for Global Environment Change
WTC	World Trade Center

Acknowledgements

The starting point of this volume was a conference on tolerability of risk held in March 2004 and funded by Pfizer Global Research and Development, the Swedish Research Foundation and the King's Risk Forum. The editors of this volume are grateful to all the individuals who participated in the conference, and especially to those who agreed to present their views, including Ben Ale, Tony Bandle, Declan Doogan, Robyn Fairman, Baruch Fischhoff, Jean-Marie Le Guen, Jim McQuaid, Jeffrey Podger, Ortwin Renn, Jacques Repussard. The editors would also like to express their special gratitude to the participants who contributed a chapter to this volume. We would also like to thank senior regulators at EMEA (European Medicines Agency), FDA (Food and Drug Administration), FSA (Food Standards Agency), HSE (Health and Safety Executive) and MHRA (Medicines and Healthcare products Regulatory Agency), who have since agreed, on various occasions, to share their views on tolerability of risk.

For the material making up Part I of this book, Ortwin Renn gratefully acknowledges the financial support for the research leading to this paper by the International Risk Governance Council (IRGC), the Swiss State Secretariat for Education and Research, the Swiss Federal Agency for Development and Cooperation, Electricité de France and the Swiss Reinsurance Company. Substantial input to this paper was provided by five background papers presented at an IRGC workshop held in fall 2004 in Ismaning sponsored by the Allianz Center for Technology in Munich. In particular the paper by Jean-Pierre Contzen on organizational capacity has been largely adopted for the section on organizational capacity building. Peter Graham provided an extensive review of existing frameworks, which inspired much of the discussion in this paper. Caroline Kuenzi carefully edited the whole manuscript, and added several paragraphs, and, with Chris Bunting, provided language editing and suggestions for case examples. Howard Kunreuther provided valuable material for the section on interpretative ambiguity and interdependencies. The members of the IRGC's CHARGOV group gave helpful advice and constructive feedback in all stages of completing the manuscript. Those members are: Lutz Cleemann, Jean-Pierre Contzen, Peter Graham, Wolfgang Kröger, Harry Kuiper, Joyce Tait and Jonathan Wiener. The author is also indebted to the members of the IRGC Scientific and Technical Council for their feedback, and particularly to Manuel Heitor, who acted as review coordinator. He is also grateful to the five anonymous reviewers of the manuscript who provided constructive criticism and suggestions for improvement. Additional reviews and input were also received from Eugene Rosa, Paul Stern, Granger Morgan, Ragnar Löfstedt, Frederic Boulder, Marion Dreyer, Juergen Hampel, Alexander Jäger, Pia-Johanna Schweizer and the participants at the above-mentioned workshop.

Introduction

Frédéric Bouder, David Slavin and Ragnar E. Löfstedt

> *The major interest groups are bought together and encouraged to conclude a series of bargains about their future behaviour, which will have the effect of moving economic events along the desired path. The plan indicates the general direction in which the interest groups, including the state in its various guises, have agreed that they want to go.* (Shonfeld, 1965)

Regulating health risks is a most serious matter, because it potentially involves pain or relief, life or death. Public debates on risks to human health are often presented along simplistic lines, by answering basic questions such as 'Are we safe enough?' or 'Are we well protected?' Answering these simple questions may often prove to be more complex than one would think.

Have health regulators done a good job so far? Many factors driven by regulatory decisions do indeed contribute to making us healthier than before. We smoke less, thanks to anti-tobacco legislation. We engage less in painful and exhausting physical work, thanks to worker-protection laws. A larger number of people have access to new treatments, thanks to effective market authorization of innovative drugs. On the other hand, it seems that regulators have neglected to address some unhealthy aspects of modern life. For example, we observe a massive increase in obesity, as well as a dramatic increase in related diseases such as diabetes. These developments may have devastating public health effects. In addition, a superficial look at how the media report regulators' aptitude to deal with health risks conveys a muddled picture of how we might assess the regulators' performance. In the pharmaceutical area, for example, the media often tell us stories about wonder drugs that will save people's lives, and new treatments that are desperately needed, calling for more rapid authorization procedures.[1] At the same time, the media seem increasingly alarmed about potential failures of the regulatory system to protect us from specific drug-related risks.[2]

Academics who have closely watched regulatory decisions, most of which are concerned in one way or another with human health, seem to give credit to the idea that regulators are confronted with a very difficult task, that allocation of risks and benefits is not simple. Sometimes they miss some important risks and sometimes they spend a lot of money and energy on dealing with negligible risks. According to David Vogel, 'numerous studies of American health and safety standards have demonstrated the inconsistency of the risks assessments that underlie them' (Vogel, 2001). Are regulatory processes doomed to criticism? Is there a way to improve our understanding of the balance of negatives and positives that each risk decision implies? What could be the gold standards of health risk regulation?

In the UK, the Health and Safety Executive (HSE), which is mainly responsible for occupational health, is using a combination of scientific and psychological approaches to help make decisions about risks that we need to refuse, risks that we can accept and risks that we may tolerate, relying on a mixture of incentives and control mechanisms:

> *The HSE's approach to the securing of compliance with risk-reducing measures is marked by the (generally consensual) use of a wide variety of techniques. Statutory duties, regulations, approved codes of practice and guidance notes are applied through prosecutions, the issue of prohibition or improvement notices, persuasion, negotiation, education, bluff, advice, information sharing and promotional work. [...] Whatever the regulatory tool or enforcement strategy, the HSE describes its general approach as one designed to manage risks by enlisting the cooperation of those affected; fostering a culture disposing those involved to give their best; planning and setting priorities for ensuring that risks requiring most attention are tackled first; setting up a system for monitoring and evaluation progress; and applying sound engineering practice.* (Baldwin et al, 2000)

The main objective of this book on the tolerability of risk (ToR) is to pool the collective knowledge of leading experts and practitioners in order to review in depth concepts and practices around notions of the acceptability and tolerability of risk. Two academics and three practitioners guide us on this stimulating journey. The book begins with a comprehensive and integrated analytic framework for risk governance by Ortwin Renn. The framework provides guidance for the development of comprehensive assessment and management strategies to cope with risks, including dealing with the complex choices involved in risk acceptance and tolerance. The next three contributions look in detail at the most sophisticated regulatory heuristics developed to date for making decisions about risk acceptability. Former or current members of the HSE present the finer points of the past, present and future of the ToR framework. Jim McQuaid contributes a unique and insightful historical perspective on ToR, focusing especially on how this concept was developed and stressing its significance in the UK risk culture. Tony Bandle gives us the regulator's perspective on a framework which, as he explains, facilitates 'risk-based regulation of work activities that [...] is all about sensible, proportionate risk management which is, itself, essentially a matter of applying sound judgement'. And Jean-Marie Le Guen looks more specifically at the implementation of the ToR framework and the current support that it may generate. In the last contribution of the book, Robyn Fairman raises the issue of 'what makes ToR work' and what the possibilities or limitations of its applicability to other risk fields might be. She argues that, for ToR to be put into operation successfully, two fundamental components are necessary: the first prerequisite is an acceptance and a legitimization by stakeholders of the need to balance risks being regulated or created against the costs involved in controlling such risks; the second is a form of institutional decision-making that allows a 'balancing of risks and costs' but ensures that decisions are reached.

Considering the impressive body of evidence in front of us (HSE, 1988, 2001), we would argue that the ToR framework has indeed been very successful at dealing with the critical issues of decisions on risk acceptability, which, Renn tells us, is a central component of sound risk governance. ToR is a dynamic concept, characterized by the integration into the decision-making process of estimates about individual and societal risks, as various perceptions may have very different tolerability patterns (Fischhoff et al, 2000; Fischhoff, 1983; Slovic, 1987). According to the HSE, 'criteria for individual risk are a necessary but cannot by themselves always be a sufficient condition. [...] Societal risk is known to have been an important, and sometimes dominant, additional issue' (HSE, 1989). This effort to combine individual with societal risk is the very essence of the ToR concept.

This balanced approach between individual and societal risks can be characterized as a negotiation process – what economists would call a trade-off – between individual risk estimates and value preferences of stakeholders. In other terms, it could also be argued that the ToR concept tries to achieve a conceptual balance between utility (the need to ensure decisions about risk are established on the basis of sufficiently broad and reliable estimates) and equity (ensuring that all social concerns are taken on board in a proportionate way). The ToR approach leads to the possibility of early mitigation measures and to a more effective marshalling of resources. It also contributes to informed decisions throughout the entire process, as the regulatee is invited to reflect on its own choices and to justify them.

The framework provides satisfaction to both the regulator and industry, as it simultaneously ensures high levels of compliance and manageable expectations On the industry side, the framework is a call for proactive self-compliance. On the regulatory side it avoids expensive enforcement and ensures higher levels of compliance than the imposition of stringent top-down duties. Risk avoidance is made difficult and action plans resulting from such measures have, in practice, been driven by a reasonable interpretation of duty to maintain risks as low as reasonably practicable. Fundamentally this approach is an aid to consistent, transparent, risk-based decision-making to communicate competence and address concerns.

Notes

1. This is, for example, the case of the recent debate on access to Herceptin in the UK (see www.guardian.co.uk/science/story/0,,1742168,00.html).
2. The *Financial Times*, for example, expressed such worries during the Vioxx crisis of 2004–2005, with headlines such as 'Master or servant: The US drugs regulator is put under scrutiny', 'Drugmakers hit by misleading ads' and 'Painkiller linked to 140,000 heart attacks in patients'.

References

Baldwin, R., Hutter, B. and Rothstein, H. (2000) 'Risk regulation, management and compliance: A report to the BRI Inquiry', London School of Economics and Political Science, London

Fischhoff, B. (1983) 'Acceptable risk: The case of nuclear power', *Journal of Policy Analysis and Management*, vol 2, pp559–575

Fischhoff, B., Slovic, P., Lichtenstein, S., Read, S. and Combs, B. (2000) 'How safe is safe enough? A psychometric study of attitudes towards technological risks and benefits', in P. Slovic (ed) *The Perception of Risk*, Earthscan, London, pp80–103

HSE (Health and Safety Executive) (1988, revised 1992) *The Tolerability of Risks from Nuclear Power Stations*, HSE Books, Sudbury

HSE (1989, revised 1994) *Quantified Risk Assessment: Its Input to Decision-Making*, HSE Books, Sudbury

HSE (2001, revised 2002) *Reducing Risks, Protecting People*, HSE Books, Sudbury

Shonfield, A. (1965) *Modern Capitalism*, Oxford University Press, Oxford

Slovic, P. (1987) 'Perception of risk', *Science*, no 236, 17 April

Vogel, D. (2001) 'The new politics of risk regulation in Europe', Discussion Paper 3, Centre for Analysis of Risk and Regulation, London School of Economics and Political Science, London

Part I

Concepts

The 'Concepts' section of this volume is divided into three chapters. Chapter 1 presents the components of the risk governance framework, Chapter 2 goes through the risks handling chain and Chapter 3 suggests looking at wider governance issues. Together, these three chapters put forward an integrated analytic framework for risk governance which provides guidance for the development of comprehensive assessment and management strategies to cope with risks. The framework integrates scientific, economic, social and cultural aspects and includes the effective engagement of stakeholders. It therefore offers a conceptual framework to guide decisions about which risks we might accept or refuse. This framework has been developed under the direction of the International Risk Governance Council (IRGC) and published as IRGC White Paper No 1, 'Risk governance: Towards an integrative approach' (2005, available at www.irgc.org/irgc/knowledge_centre/irgcpublications/).

The concept of risk governance comprises a broad picture of risk: not only does it include what has been termed 'risk management' or 'risk analysis', it also looks at how risk-related decision-making unfolds when a range of actors are involved, requiring coordination and possibly reconciliation between a profusion of roles, perspectives, goals and activities. Indeed the problem-solving capacities of individual actors, be they government, the scientific community, business players, NGOs or civil society as a whole, are limited and often unequal to the major challenges facing society today. Risks such as those related to increasingly violent natural disasters, food safety or critical infrastructures call for coordinated effort among a variety of players beyond the frontiers of countries, sectors, hierarchical levels, disciplines or risk fields. Finally, risk governance also illuminates a risk's context by taking account of such factors as the historical and legal background, guiding principles, value systems and perceptions, and organizational imperatives.

1

Components of the Risk Governance Framework

Ortwin Renn

The risk governance framework offers two major innovations to the field of risk: the inclusion of the societal context and a new categorization of risk-related knowledge.

1   *Inclusion of the societal context.* Besides the generic elements of risk assessment, risk management and risk communication, the framework gives equal importance to contextual aspects, which are either directly integrated in a model risk process comprising of the above and additional elements or form the basic conditions for making any risk-related decision. Contextual aspects of the first category include the structure and interplay of the different actors dealing with risks, how these actors may differently perceive the risks and what concerns they have regarding their likely consequences. Examples of the second category include the policy-making or regulatory style as well as the socio-political impacts prevalent within the entities and institutions having a role in the risk process, their organizational imperatives and the capacity needed for effective risk governance. Linking the context with risk governance, the framework reflects the important role of risk–benefit evaluation and the need for resolving risk–risk trade-offs.

2   *Categorization of risk-related knowledge.* The framework also proposes a categorization of risk which is based on the different states of knowledge about each particular risk, distinguishing between 'simple', 'complex', 'uncertain' and 'ambiguous' risk problems. The characterization of a particular risk depends on the degree of difficulty of establishing the cause–effect relationship between a risk agent and its potential consequences, the reliability of this relationship and the degree of controversy with regard to both what a risk actually means for those affected and the values to be applied when judging whether or not something needs to be

done about it. Examples of each risk category include, respectively, known health risks such as those related to smoking; the failure risk of interconnected technical systems such as the electricity transmission grid; atrocities such as those resulting from the changed nature and scale of international terrorism; and the long-term effects and ethical acceptability of controversial technologies such as nanotechnologies. For each category, a strategy is then derived for risk assessment and risk management, as well as the level and form of stakeholder participation, supported by proposals for appropriate methods and tools.

Beyond these component parts the framework includes three major value-based premises and assumptions:

1   The framework is inspired by the conviction that both the 'factual' and the 'socio-cultural' dimension of risk need to be considered if risk governance is to produce adequate decisions and results. While the factual dimension comprises physically measurable outcomes and discusses risk in terms of a combination of potential consequences – both positive and negative – and the probability of their occurrence, the socio-cultural dimension emphasizes how a particular risk is viewed when values and emotions come into play.

2   The inclusiveness of the governance process: regarded as a necessary (although not in itself sufficient) prerequisite for tackling risks in both a sustainable and acceptable manner, it consequently imposes an obligation to ensure the early and meaningful involvement of all stakeholders, civil society in particular.

3   The framework implements the principles of good governance: beyond the crucial commitment to participation these principles include transparency, effectiveness and efficiency, accountability, strategic focus, sustainability, equity and fairness, respect for the rule of law, and the need for the chosen solution to be politically and legally realizable as well as ethically and publicly acceptable.

The Core of the Framework: The Risk Governance Phases

The framework's risk process, or risk handling chain, is illustrated in Figure 1.1. It breaks down into three main phases: 'pre-assessment', 'appraisal' and 'management'. A further phase, comprising the 'characterization' and 'evaluation' of risk, is placed between the appraisal and management phases and can be assigned to either those charged with the assessment or those responsible for management, depending on who are better equipped to perform the associated tasks, thus concluding the appraisal phase or marking the start of the management phase. The risk process has 'communication' as a companion to all phases

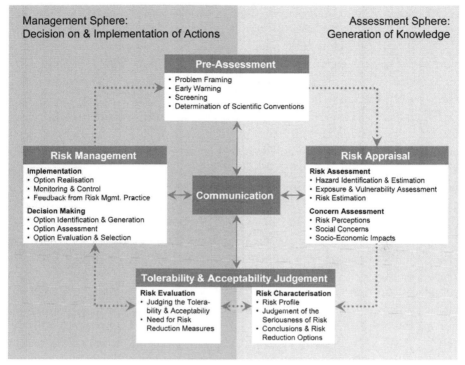

Figure 1.1 The risk governance framework

of addressing and handling risk and is itself of a cyclical nature. However, the clear sequence of phases and steps offered by this process is primarily a logical and functional one and will not always correspond to reality.

The purpose of the pre-assessment phase is to capture both the variety of issues that stakeholders and society may associate with a certain risk and existing indicators, routines and conventions that may prematurely narrow down, or act as a filter for, what is going to be addressed as risk. What counts as a risk may be different for different groups of actors. The first step of pre-assessment – risk framing – therefore places particular importance on the need for all interested parties to share a common understanding of the risk issues being addressed or raising awareness among those parties of the differences in what is perceived as a risk. For a common understanding to be achieved, actors need both to agree with the underlying goal of the activity or event generating the risk and be willing to accept the risk's foreseeable implications for that very goal. A second step of the pre-assessment phase – early warning and monitoring – establishes whether signals of the risk exist that might indicate its realization. This step also investigates the institutional means in place for monitoring the environment for such early-warning signals. The third step –

pre-screening – takes up and looks into the widespread practice of conducting preliminary probes into hazards or risks and, based on prioritization schemes and existing models for dealing with risk, of assigning a risk to pre-defined assessment and management 'routes'. The fourth and final step of pre-assessment selects major assumptions, conventions and procedural rules for assessing both the risk and the emotions associated with it.

The objective of the risk appraisal phase is to provide the knowledge base for a societal decision on whether or not a risk should be taken and, if so, how it might be reduced or contained. Risk appraisal thus comprises a scientific assessment of both the risk and questions that stakeholders may have concerning its social and economic implications.

The first component of risk appraisal – risk assessment – seeks to link a potential source of harm, a hazard, with likely consequences, specifying probabilities of occurrence for the latter. Depending on the source of a risk and the organizational culture of the community dealing with it, many different ways exist for structuring risk assessment. Despite such diversity, however, three core steps can be identified: the identification and, if possible, estimation of the hazard, an assessment of related exposure and/or vulnerability, and an estimation of the consequent risk. The last step – risk estimation – aggregates the results of the first two steps and gives a probability of occurrence for each conceivable degree of severity of the consequences. Confirming the results of risk assessments can be extremely difficult, in particular when cause–effect relationships are hard to establish, when they are unstable due to variations in both causes and effects, and when effects are both scarce and difficult to understand. Depending on the achievable state and quality of knowledge, risk assessment is thus confronted with three major challenges that can best be summarized using the risk categories outlined above – 'complexity', 'uncertainty' and 'ambiguity'. For a successful outcome to the risk process and, indeed, overall risk governance, it is crucial that the implications of these challenges are made transparent at the conclusion of risk assessment and throughout all subsequent phases.

Equally important to understanding the physical attributes of the risk is detailed knowledge of stakeholders' concerns and questions – emotions, hopes, fears and apprehensions – about the risk as well as likely social consequences, economic implications and political responses. The second component of risk appraisal – concern assessment – thus complements the results from risk assessment with insights from risk perception studies and interdisciplinary analyses of the risk's (secondary) social and economic implications.

The most controversial phase of handling risk – risk characterization and evaluation – aims at judging a risk's acceptability and/or tolerability. A risk deemed 'acceptable' is usually limited in terms of negative consequences so is taken on without risk reduction or mitigation measures being envisaged. A risk deemed 'tolerable' links undertaking an activity which is considered worthwhile for the value-added or benefit it provides with specific measures to diminish and limit the likely adverse consequences. This judgement is informed by two distinct but closely related efforts to gather and compile the necessary knowledge which, in the case of tolerability, must additionally support an initial understanding of required risk reduction and mitigation measures. While risk characterization compiles scientific evidence based on the results from the risk appraisal phase, risk evaluation assesses broader value-based issues that also influence the judgement. Such issues, which include questions such as the choice of technology, societal needs requiring a given risk agent to be present, and the potential for substitution as well as for compensation, reach beyond the risk itself and into the realm of policy-making and societal balancing of risks and benefits.

The risk management phase designs and implements the actions and remedies required to tackle risks with the aim of avoiding, reducing, transferring or retaining them. Risk management thereby relies on a sequence of six steps which facilitates systematic decision-making. To start with, and based on a reconsideration of the knowledge gained in the risk appraisal phase and while judging the acceptability and/or tolerability of a given risk, a range of potential risk management options is identified. These options are then assessed with regard to criteria such as effectiveness, efficiency, minimization of external side effects, sustainability and so forth. These assessment results are next complemented by a value judgement on the relative weight of each of the assessment criteria, allowing an evaluation of the risk management options. This evaluation supports the next step in which one (or more) of the risk management options is selected, normally after consideration of possible trade-offs that need to be made between a number of second-best options. The final two steps include the implementation of the selected options and the periodic monitoring and review of their performance.

Based on the dominant characteristic of each of the four risk categories ('simple', 'complex', 'uncertain', 'ambiguous') it is possible to identify specific safety principles and, consequently, design a targeted risk management strategy (see Table 1.1). 'Simple' risk problems can be managed using a 'routine-based' strategy which draws on traditional decision-making instruments, best practice as well as time-honoured trial and error. For 'complex' and 'uncertain' risk problems it is helpful to distinguish the strategies required to deal with a risk agent from those directed at the risk-absorbing system: complex risks are thus usefully addressed on the basis of 'risk-informed' and 'robustness-focused' strategies, while uncertain risks are better managed using 'precaution-based' and 'resilience-focused' strategies. Whereas the former strategies aim at accessing and acting on the best available scientific expertise and at reducing a system's vulnerability to known hazards and threats by improving its buffer capacity, the latter strategies pursue the goal of applying a precautionary approach in order to

Table 1.1 *Risk characteristics and their implications for risk management*

Knowledge Characterization	Management Strategy
1 'Simple' risk problems	*Routine-based:* (tolerability/acceptability judgement) (risk reduction)
2 Complexity-induced risk problems	*Risk-informed:* (risk agent and causal chain) *Robustness-focused:* (risk-absorbing system)
3 Uncertainty-induced risk problems	*Precaution-based:* (risk agent) *Resilience-focused:* (risk-absorbing system)
4 Ambiguity-induced risk problems	*Discourse-based:*

ensure the reversibility of critical decisions and of increasing a system's coping capacity to the point where it can withstand surprises. Finally, for 'ambiguous' risk problems the appropriate strategy consists of a 'discourse-based' strategy which seeks to create tolerance and mutual understanding of conflicting views and values with a view to eventually reconciling them.

Appropriate Instruments	Stakeholder Participation
• Applying 'traditional' decision-making – Risk–benefit analysis – Risk–risk trade-offs – Trial and error – Technical standards – Economic incentives – Education, labelling, information – Voluntary agreements	Instrumental discourse

• Characterizing the available evidence
– Expert consensus-seeking tools:
  – Delphi or consensus conferencing
  – Meta-analysis
  – Scenario construction, etc
– Results fed into routine operation
 	
• Improving buffer capacity of risk target through:
– Additional safety factors
– Redundancy and diversity in designing safety devices
– Improving coping capacity
– Establishing high reliability organizations
	Epistemological discourse

• Using hazard characteristics such as persistence, ubiquity, etc as proxies for risk estimates Tools include: – Containment – ALARA (as low as reasonably achievable) and ALARP (as low as reasonably possible) – BACT (best available control technology), etc • Improving capability to cope with surprises – Diversity of means to accomplish desired benefits – Avoiding high vulnerability – Allowing for flexible responses – Preparedness for adaptation	Reflective discourse
• Application of conflict resolution methods for reaching consensus or tolerance for risk evaluation results and management option selection – Integration of stakeholder involvement in reaching closure – Emphasis on communication and social discourse	Participative discourse

The remaining element of the risk process is risk communication, which is of major importance throughout the entire risk handling chain. Not only should risk communication enable stakeholders and civil society to understand the rationale of the results and decisions from the risk appraisal and risk management phases when they are not formally part of the process, it should also help them to make informed choices about risk, balancing factual knowledge about risk with personal interests, concerns, beliefs and resources when they are themselves involved in risk-related decision-making. Effective risk communication consequently fosters tolerance for conflicting viewpoints, provides the basis for their resolution, and creates trust in the institutional means for assessing and managing risk and related concerns. Eventually, risk communication can have a major impact on how well society is prepared to cope with risk and react to crises and disasters. Risk communication has to perform these functions both for the experts involved in the overall risk process – requiring the exchange of information between risk assessors and managers, between scientists and policymakers, between academic disciplines, and across institutional barriers – and for the 'outside world' of those affected by the process.

Wider Governance Issues: Organizational Capacity and Regulatory Styles

This chapter also addresses wider governance issues pertinent to the context of a risk and the overall risk process, thus acknowledging the many different pathways that different countries or, indeed, risk communities, may pursue for dealing with risk. The discussion of these wider issues begins with an assessment of the very notion of 'risk governance' which builds on the observation that collective decisions about risks are the outcome of a 'mosaic' of interactions between governmental or administrative actors, scientific communities, corporate actors and actors from civil society at large, many of the interactions taking place and relevant to only individual parts of the overall process. The interplay of these actors has various dimensions, including public participation, stakeholder involvement and the formal (horizontal and vertical) structures within which it occurs. This section on concepts additionally investigates organizational prerequisites for effective risk governance, which are at the crossroads of the formal responsibilities of actors and their capability and authority to successfully fulfil their roles, and makes a very short case for risk education. The organizational prerequisites are summarized under 'institutional and organizational capacity' and include both intellectual and material 'assets' and 'skills' and the framework of relations, or 'capabilities', required to make use of these. The discussion of wider risk governance issues concludes with a reflection on the role of political culture and a proposal for a typology of different regulatory regimes or governmental styles.

Scope of the Proposed Framework

This section covers a wide range of both risks and governance structures. 'Risk' is understood as an uncertain consequence of an event or an activity with respect to something that humans value, a definition originally used by Kates et al (1985, p21). Risks always refer to a combination of two components: the likelihood or chance of potential consequences and the severity of consequences of human activities, natural events or a combination of the two. Such consequences can be positive or negative, depending on the values that people associate with them. In addition to the strength and likelihood of these consequences, the framework emphasizes the distribution of risks over time, space and populations. In particular, the timescale of appearance of adverse effects is very important and links risk governance to sustainable development ('delayed effects').

In this section we distinguish risks from hazards. 'Hazards' describe the potential for harm or other consequences of interest. These potentials may never actually materialize, however, if, for example, people are not exposed to the hazards or if the targets are made resilient against the hazardous effect (such as immunization). In conceptual terms, hazards characterize the *inherent properties of the risk agent and related processes*, whereas risks describe the *potential effects that these hazards are likely to cause on specific targets such as buildings, ecosystems or human organisms and their related probabilities.*

Risk in a Broader Context

The focus on risk should be seen as a segment of a larger and wider perspective on how humans transform the natural into a cultural environment with the aims of improving living conditions and serving human wants and needs (Turner et al, 1990). These transformations are performed with a purpose in mind (normally a benefit to those who initiate them). When implementing these changes, intended (or tolerated) and unintended consequences may occur that meet or violate other dimensions of what humans value. Risks are not taken for their own sake; rather they are actively or passively incurred because of their being an integral factor in the very activity that is geared towards achieving the particular human need or purpose. In this context it is the major task of risk assessment to identify and explore, preferably in quantitative terms, the types, intensities and likelihood of the normally undesired consequences related to risks. In addition, these consequences are associated with special concerns that individuals, social groups or different cultures may attribute to such risks. They also need to be assessed in order to make a prudent judgement about the tolerability or acceptability of risks. Once this judgement is made it is the task of risk management to prevent, reduce or alter these consequences by choosing appropriate actions. As obvious as this distinction between risk and concern assessment (as a tool for gaining knowledge about risks) and risk management (as a tool for handling risks) appears at first glance, however, the distinction becomes blurred in the actual risk governance process.

This blurring is due to the fact that assessment starts with the respective risk agent or source and tries both to identify potential damage scenarios and their probabilities and to model its potential consequences over time and space, whereas risk management oversees a much larger terrain of potential interventions (Stern and Fineberg, 1996; Jasanoff, 1986, p79ff; Jasanoff, 2004). Risk management may alter human wants or needs (so that the agent is not even created or continued); it can suggest substitutes or alternatives for the same need; it can relocate or isolate activities so that exposure is prevented; or it can make risk targets less vulnerable to potential harm. Figure 1.2 illustrates this larger perspective for technological risks and lists the possible intervention points for risk management. Risk assessment and management are therefore not symmetrical to each other: management encompasses a much larger domain and may even occur before assessments are performed. It is, moreover, often based on considerations that are not affected by or part of the assessment results. In more general terms, risk management refers to the creation and evaluation of options for initiating or changing human activities or natural and artificial structures with the objective being to increase the net benefit to human society and prevent harm to humans and what they value. The identification of these options and their evaluation is guided by systematic and experiential knowledge gained and prepared for this purpose by experts and stakeholders. A major proportion of that relevant knowledge comprises the results of risk assessments. However, risk managers also need to act in situations of 'non-knowledge' or insufficient knowledge about potential outcomes of human actions or activities.

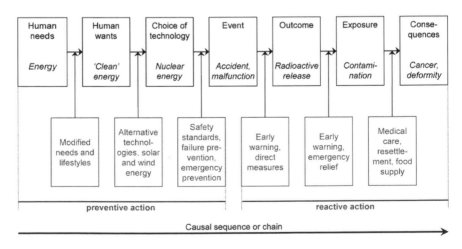

Figure 1.2 Seven steps of a risk chain: The example of nuclear energy

Source: Hohenemser et al, 1983

The most complex questions emerge, however, when one looks at how society and its various actors actually handle risk. In addition to knowledge gained through risk assessments and/or option generation and evaluation through risk management, the decision-making structure of a society is itself highly complicated and often fragmented. Apart from the structure itself – the people and organizations that share responsibility for assessing and managing risk – one must also consider the need for sufficient organizational capacity to create the necessary knowledge and implement the required actions, the political and cultural norms, rules and values within a particular societal context, and the subjective perceptions of individuals and groups. These factors leave their marks on the way risks are treated in different domains and socio-political cultures.

In the last decade the term 'governance' has experienced tremendous popularity in the literature on international relations, comparative political science, policy studies, sociology of environment and technology, as well as risk research.[1] On a national scale, governance describes *structures and processes for collective decision making involving governmental and non-governmental actors* (Nye and Donahue, 2000). Governing choices in modern societies is seen as an interplay between governmental institutions, economic forces and civil society actors (such as NGOs). At the global level, governance *embodies a horizontally organized structure of functional self-regulation encompassing state and non-state actors bringing about collectively binding decisions without superior authority* (Rosenau, 1992; Wolf, 2002). In this perspective non-state actors play an increasingly relevant role and become more important, since they have decisive advantages in terms of information and resources compared to single states.

It is also useful to differentiate between 'horizontal' and 'vertical' governance (Benz and Eberlein, 1999; Lyall and Tait, 2004). The horizontal level includes the relevant actors in decision-making processes within a defined geographical or functional segment (such as all relevant actors within a community, region, nation or continent); the vertical level describes the links between these segments (such as the institutional relationships between the local, regional and state levels).

'Risk governance' involves the 'translation' of the substance and core principles of governance to the context of risk and risk-related decision-making. It includes the totality of actors, rules, conventions, processes and mechanisms concerned with how relevant risk information is collected, analysed and communicated and management decisions are taken. Encompassing the combined risk-relevant decisions and actions of both governmental and private actors, risk governance is of particular importance in, but not restricted to, situations where there is no single authority to take a binding risk-management decision but where, instead, the nature of the risk requires the collaboration of, and coordination between, a range of different stakeholders. Risk governance not only includes a multifaceted, multi-actor risk process, however; it also calls for the consideration of contextual factors such as institutional arrangements (for example the regulatory and legal framework that determines the relationship, roles and responsibilities of the actors and coordination mechanisms such as markets, incentives or self-imposed norms) and political culture, including different perceptions of risk.

When looking at risk governance structures there is no possibility of including all the variables that may influence the decision-making process: there are simply too many. Therefore it is necessary to limit one's efforts to those factors and actors that, by theoretical reasoning and/or empirical analysis, are demonstrably of particular importance with respect to the outcome of risk governance. For example:

- the structure and function of various actor groups in initiating, influencing, criticizing and/or implementing risk policies and decisions;
- risk perceptions of individuals and groups;
- individual, social and cultural concerns associated with the consequences of risk;
- the regulatory and decision-making style (political culture); and
- the requirements with respect to organizational and institutional capabilities for assessing, monitoring and managing risks (including emergency management).

In addition to these analytical categories, this volume also addresses best practice and normative aspects of what is needed to improve governance structures and processes (European Commission, 2001). With respect to best practice it is interesting to note that often risk creators, in particular when directly affected by the risk they generate, engage in risk reduction and avoidance out of self-interest or on a voluntary basis (for example industrial 'gentlemen's agreements', self-restriction and industry standards). Other stakeholders' efforts in risk governance therefore have to be coordinated with what is tacitly in place already. The emphasis here is on cooperative models of public–private partnerships forming a governance system that aims at effective, efficient and fair risk-management solutions.[2]

Notes

1   According to Rhodes (1996) there are six separate uses of the term governance: as minimal state, as corporate governance, as new public management, as good governance, as social-cybernetic systems and as self-organized networks.
2   Excluded from this volume are topics such as crisis intervention, crisis communication, emergency planning and management, and post-accidental relief. These will be covered in a separate volume at a later stage.

References

Benz, A. and Eberlein, B. (1999) 'The Europeanization of regional policies: Patterns of multi-level governance', *Journal of European Public Policy*, vol 6, no 2, pp329–348

European Commission (2001) *European Governance: A White Paper*, EU, Brussels

Hohenemser, C., Kates, R. W. and Slovic, P. (1983) 'The nature of technological hazard', *Science*, no 220, pp378–384

Jasanoff, S. (1986) *Risk Management and Political Culture*, Russell Sage Foundation, New York

Jasanoff, S. (2004) 'Ordering knowledge, ordering society', in S. Jasanoff (ed) *States of Knowledge: The Co-Production of Science and Social Order*, Routledge, London, pp31–54

Kates, R. W., Hohenemser, C. and Kasperson, J. (1985) *Perilous Progress: Managing the Hazards of Technology*, Westview Press, Boulder, CO

Lyall, C. and Tait, J. (2004) 'Shifting policy debates and the implications for governance', in C. Lyall and J. Tait (eds) *New Modes of Governance: Developing an Integrated Policy Approach to Science, Technology, Risk and the Environment*, Ashgate, Aldershot, UK, pp3–17

Nye, J. S. and Donahue, J. (eds) (2000) *Governance in a Globalising World*, Brookings Institution, Washington

Rhodes, R. A. W. (1996) 'The new government: Governing without government', *Political Studies*, vol 44, pp652–667

Rosenau, J. N. (1992) 'Governance, order, and change in world politics', in J. N. Rosenau and E. O. Czempiel (eds) *Governance without Government: Order and Change in World Politics*, Cambridge University Press, Cambridge, pp1–29

Stern, P. C. and Fineberg, V. (1996) *Understanding Risk: Informing Decisions in a Democratic Society*, National Research Council, Committee on Risk Characterization, National Academy Press, Washington

Turner, B. L., Clark, W. C., Kates, R. W., Richards, J. F., Mathews, J. T. and Meyer, W. B. (1990) *The Earth as Transformed by Human Action*, Cambridge University Press, Cambridge

Wolf, K. D. (2002) 'Contextualizing normative standards for legitimate governance beyond the state', in J. R. Grote and B. Gbikpi (eds) *Participatory Governance: Political and Societal Implications*, Leske und Budrich, Opladen, Germany, pp35–50

2

The Risk Handling Chain

Ortwin Renn

Risks are mental 'constructions' (OECD, 2003, p67). They are not real phenomena but originate in the human mind. Actors, however, creatively arrange and reassemble signals that they get from the 'real world', providing structure and guidance to an ongoing process of reality enactment.[1] Thus risks represent what people observe in reality and what they experience. The link between risk as a mental concept and reality is forged through the experience of actual harm (the consequence of risk) in the sense that human lives are lost, health impacts can be observed, the environment is damaged or buildings collapse. The invention of risk as a mental construct is contingent on the belief that human action can prevent harm in advance. Humans have the ability to design different futures; in other words they construct scenarios that serve as tools for the human mind to anticipate consequences in advance and change, within the constraints of nature and culture, the course of actions accordingly.

The status of risk as a mental construct has major implications for how risk is looked at. Unlike trees or houses, one cannot scan the environment, identify the objects of interest, and count them. Risks are created and selected by human actors. What counts as a risk to someone may be an act of God, or even an opportunity, for someone else. Although over time societies have gained experience and collective knowledge of the potential impacts of events and activities, one cannot anticipate all potential scenarios and be worried about all the many potential consequences of a proposed activity or an expected event. By the same token, it is impossible to include all possible options for intervention. Therefore societies have been *selective* in what they have chosen to be worth considering and what to ignore (Thompson et al, 1990; Douglas, 1990; Beck, 1994, p9ff). Specialized organizations have been established to monitor the environment for hints of future problems and to provide early warning of some potential future harm. This selection process is not arbitrary; it is guided by cultural values (such as the shared belief that each individual life is worth protecting), by institutional and financial resources (such as the decision of national governments to spend money or not to spend money on early-warning systems against highly improbable but high-consequence events) and by systematic reasoning (such as using probability theory in distinguishing between more likely and less likely events or methods to estimate damage potential or distribution of hazards in time and space).

Based on these preliminary thoughts, a systematic review of risk-related actions needs to start with an analysis of what major societal actors such as governments, companies, the scientific community and the general public select as risks and what types of problems they label as risk problems (rather than opportunities or innovation potentials, for example). In technical terms this is called 'framing'. Framing in this context encompasses the selection and interpretation of phenomena as relevant risk topics (Tversky and Kahneman, 1981; van der Sluijs et al, 2003; Goodwin and Wright, 2004). This process of framing is already part of governance structure since official agencies (for example food standard agencies), risk and opportunity producers (such as the food industry), those affected by risks and opportunities (such as consumer organizations) and interested bystanders (such as the media or an intellectual elite) are all involved and often in conflict with each other when framing the issue. What counts as risk may vary among these actor groups. Consumers may feel that all artificial food additives pose a risk, whereas industry may be concerned about pathogens that develop their negative potential due to the lack of consumer knowledge about food storage and preparation. Environmental groups may be concerned about the risks of industrial food versus organic food. Whether a consensus evolves about what requires consideration as a relevant risk depends on the legitimacy of the selection rule. The acceptance of selection rules rests on two conditions: first, all actors need to agree with the *underlying goal* (often legally prescribed, such as prevention of health detriments or guarantees of undisturbed environmental quality, for example purity laws for drinking water); second, they need to agree with the *implications derived from the present state of knowledge* (whether and to what degree the identified hazard impacts the desired goal). Even within this preliminary analysis, dissent can result from conflicting values as well as conflicting evidence, and especially from the inadequate blending of the two. Values and evidence can be viewed as the two sides of a coin: the values govern the selection of the goal whereas the evidence governs the selection of cause–effect claims. Both need to be properly investigated when analysing risk governance, but it is of particular importance to understand the values shaping the interests, perceptions and concerns of the different stakeholders as well as to identify methods for capturing how these concerns are likely to influence, or impact on, the debate about a particular risk. The actual measurements of these impacts should then be done in the most professional manner, including the characterization of uncertainties (Keeney, 1992; Pidgeon and Gregory, 2004; Gregory, 2004).

A second part of the pre-assessment phase concerns the institutional means of *early warning and monitoring*. Even if there is a common agreement of what should be framed as risk issues, there may be problems in monitoring the environment for signals of risks. This is often due to a lack of institutional efforts to collect and interpret signs of risk and deficiencies in communication between those looking for early signs and those acting upon them. The recent tsunami catastrophe in Asia provides a more than telling example of the discrepancy between the possibility of early-warning capabilities and the decision to install or use them. It is therefore important to look at early-warning and monitoring activities when investigating risk governance.

In many risk governance processes, information about risks is pre-screened and then allocated to different assessment and management routes. In particular, industrial risk managers search for the most efficient strategy to deal with risks. This includes prioritization policies, protocols for dealing with similar causes of risks, and optimal models combining risk reduction and insurance. Public risk regulators often use pre-screening activities to allocate risks to different agencies or to predefined procedures. Sometimes risks may seem to be less severe and it may be adequate to cut short risk or concern assessment. On the other hand, in a pending crisis situation, risk management actions may need to be taken before any assessment is even carried out. A full analysis should therefore include provisions for *risk screening* and the selection of different routes for risk assessment, concern assessment and risk management. This aspect has been called 'risk assessment policy' in the Codex Alimentarius.[2] It is meant to guide the assessment process in terms of assessment and management protocols, methods of investigation, statistical procedures and other scientific conventions used in assessing risks or selecting risk reduction options. A screening process may also be employed when characterizing risks according to complexity, uncertainty and ambiguity, as we will explain later.

Another major component of pre-assessment is the *selection of conventions and procedural rules* needed for a comprehensive scientific appraisal of the risk, in other words for assessing the risk and the concerns related to it (see below). Any such assessment is based on prior informed yet subjective judgements or conventions articulated by the scientific community or a joint body of risk assessors and managers. Those judgements refer to:

- the social definition of what is to be regarded as adverse, for example by defining the 'no adverse effect level' (NOAEL) in food;
- the selection rule determining which potentially negative effects should be considered in the risk governance process, taking into consideration the fact that an infinite number of potential negative outcomes can be theoretically connected with almost any substance, activity or event;
- the aggregation rule specifying how to combine various effects within a

one-dimensional scale, for example early fatalities, late fatalities, cancer, chronic diseases and so on;
- the selection of the testing and detection methods which are presently used in risk assessment, for example the use of genomics for calculating risk from transgenic plants;
- the selection of valid and reliable methods for measuring perceptions and concerns;
- determining models to extrapolate high-dose effects to low-dose situations, for example linear, quadro-linear, exponential or other functions or assumptions about thresholds or non-thresholds in dose–response relationships;
- the transfer of animal data to humans;
- assumptions about exposure or definition of target groups; and
- the handling of distributional effects, which may cover inter-individual, inter-group, regional, social, time-related and inter-generational aspects. (Pinkau and Renn, 1998; van den Sluijs et al, 2004, p54ff)

These judgements reflect the consensus among the experts or are common products of risk assessment and management (for example from licensing special testing methods). Their incorporation in guiding scientific analyses is unavoidable and does not discredit the validity of the results. Yet it is essential that risk managers and interested parties are informed about these conventions and understand their rationale. On the one hand knowledge about these conventions can lead to a more cautious apprehension of what the assessments mean and imply, on the other they can convey a better understanding of the constraints and conditions under which the results of the various assessments hold true.

In summary, Table 2.1 provides a brief overview of the four components of pre-assessment. The table also lists some indicators that may be useful as heuristic tools when investigating different risk governance processes. The choice of indicators is not exhaustive and will vary depending on risk source and risk target, but listing the indicators serves the purpose of illustrating the types of information needed to perform the task described in each step. The title 'pre-assessment' does not mean that these steps are always taken before assessments are performed. Rather they are logically located in the forefront of assessment and management. They should also not be seen as sequential steps but as elements that are closely interlinked. Indeed, depending on the situation, early warning might precede problem framing.[3]

Risk Assessment

The purpose of risk assessment is the generation of knowledge linking specific risk agents with uncertain but possible consequences (Lave, 1987; Graham and

Table 2.1 *Components of pre-assessment in handling risks*

Pre-assessment components	Definition	Indicators
1 Problem framing	Different perspectives of how to conceptualize the issue	• dissent or consent on goals of selection rule • dissent or consent on relevance of evidence • choice of frame (risk, opportunity, fate)
2 Early warning	Systematic search for new hazards	• unusual events or phenomena • systematic comparison between modelled and observed phenomena • novel activities or events

3 Screening (risk assessment and concern assessment policy)	Establishing a procedure for screening hazards and risks and determining assessment and management route	• screening in place?
		• criteria for screening:
		  – hazard potential
		  – persistence
		  – ubiquity etc
		• criteria for selecting risk assessment procedures for:
		  – known risks
		  – emergencies etc
		• criteria for identifying and measuring social concerns

4 Scientific conventions for risk assessment and concern assessment	Determining the assumptions and parameters of scientific modelling and evaluating methods and procedures for assessing risks and concerns	• definition of no adverse effect levels (NOAEL) • validity of methods and techniques for risk assessments • methodological rules for assessing concerns

Rhomberg, 1996). The final product of risk assessment is an estimation of the risk in terms of a probability distribution of the modelled consequences (drawing on either discrete events or continuous loss functions). The different stages of risk assessment vary from risk source to risk source. Many efforts have been made to produce a harmonized set of terms and conceptual phase-model to cover a wide range of risks and risk domains (cf. Codex Alimentarius, 2001; National Research Council, 1982 and 1983; Stern and Fineberg, 1996; European Commission, 2000 and 2003). The most recent example is the risk guidance book by the International Programme on Chemical Safety (IPCS) and World Health Organization (WHO) (IPCS and WHO, 2004). Although there are clear differences in structuring the assessment process depending on risk source and organizational culture, there is agreement on basically three core components of risk assessment:

1   identification and, if possible, estimation of hazard;
2   assessment of exposure and/or vulnerability; and
3   estimation of risk, combining the likelihood and the severity of the targeted consequences, based on the identified hazardous characteristics and the exposure/vulnerability assessment.

As we have seen before it is crucial to distinguish between hazards and risks. Thus *identification* (establishing cause–effect link) and *estimation* (determining the strength of the cause–effect link) need to be performed for hazards and risks separately. The estimation of risk depends on an *exposure* and/or *vulnerability* assessment. Exposure refers to the contact of the hazardous agent with the target (individuals, ecosystems, buildings and so forth); vulnerability describes the various degrees to which the target experiences harm or damage as a result of the exposure (for example the immune systems among a target population, vulnerable groups or structural deficiencies in buildings). In many cases it is common practice to combine hazard and risk estimates in scenarios that allow modellers to change parameters and include different sets of context constraints.

The basis of risk assessment is the systematic use of analytical – largely probability-based – methods, which have been constantly improved over recent years. Probabilistic risk assessments for large technological systems, for instance, include tools such as fault and event trees, scenario techniques, distribution models based on Geographic Information Systems (GIS), transportation modelling and empirically driven human–machine interface simulations (IAEA, 1995; Stricoff, 1995). With respect to human health, improved methods for modelling individual variation (Hattis, 2004), dose–response relationships (Olin et al, 1995) and exposure assessments (US-EPA, 1997) have been developed and successfully applied. The processing of data is often guided by inferential statistics and organized in line with decision analytic procedures. These tools have been developed to generate knowledge about cause–effect relationships, estimate the strength of these relationships, characterize remaining uncertainties and ambiguities, and describe, in quantitative or qualitative form, other risk- or hazard-related properties that are important for risk management (IAEA, 1995; IEC, 1993). In short, risk assessments specify what is at stake, calculate the probabilities for wanted or unwanted consequences, and aggregate both components into a single dimension (Kolluru, 1995). In general there are five methods for calculating probabilities:

1   Collection of statistical data relating to the performance of a risk source in the past (actuarial extrapolation).

2   Collection of statistical data relating to components of a hazardous agent or technology. This method requires a synthesis of probability judgements from component failure to system performance – probabilistic risk assessments (PRAs).

3   Epidemiological or experimental studies which are aimed at finding statistically significant correlations between an exposure to a hazardous agent and an adverse effect in a defined population sample (probabilistic modelling).

4   Experts' or decision-makers' best estimates of probabilities, in particular for events where only insufficient statistical data is available (normally employing Bayesian statistical tools).

5   Scenario techniques by which different plausible pathways from release of a harmful agent to the final loss are modelled on the basis of worst and best cases or estimated likelihood of each consequence at each knot.

All these methods are based either on the past performance of the same or a similar risk source or an experimental intervention. However, the possibility that the circumstances of the risk situation may vary over time in an unforeseeable way and that people will thus make decisions in relation to changing hazards – sometimes they may even change in an unsystematic, unpredictable manner – leads to unresolved or remaining uncertainty (second order uncertainty). One of the main challenges of risk assessment is the systematic characterization of these remaining uncertainties. They can partly be modelled by using inferential statistics (confidence interval) or other simulation methods (such as Monte Carlo), but often they can only be described in qualitative terms. Risk analysts consequently distinguish between *aleatory* and *epistemic* uncertainty: epistemic uncertainty can be reduced by more scientific research,[4] while aleatory uncertainty will remain fuzzy regardless of how much research is invested in the subject (Shome et al, 1998). Remaining uncertainties pose major problems in the later stages of risk characterization and evaluation as well as risk management since they are difficult to integrate in formal risk–benefit analyses or in setting standards.

There is no doubt that risk assessment methods have matured to become sophisticated and powerful tools in coping with the potential harm of human actions or natural events (Morgan, 1990). Its worldwide application in dealing and managing risks, however, largely fails to reflect this degree of power and professionalism. And at the same time there are new challenges in the risk field that need to be addressed by the risk assessment communities. These challenges refer to:

*   widening the scope of effects for risk assessment, including chronic diseases (rather than focusing only on fatal diseases such as cancer or heart attack), risks to ecosystem stability (rather than focusing on a single

species), and the secondary and tertiary risk impacts that are associated with the primary physical risks;

- addressing risk at a more aggregated and integrated level, such as studying synergistic effects of several toxins or constructing a risk profile over a geographic area that encompasses several risk causing facilities;
- studying the variations among different populations, races and individuals and getting a more adequate picture of the ranges of sensibilities with respect to environmental pollutants, lifestyle factors, stress levels and impacts of noise;
- integrating risk assessment in a comprehensive technology assessment or option appraisal so that the practical value of its information can be phased into the decision-making process at the needed time and that its inherent limitations can be compensated through additional methods of data collection and interpretation; and
- developing more forgiving technologies that tolerate a large range of human error and provide sufficient time for initiating counteractions. (Brown and Goble, 1990; Hattis and Kennedy, 1990; Greeno and Wilson, 1995; Renn, 1997)

Table 2.2 lists the three generic components of risk assessment and provides an explanation for the terms as well as a summary list of indicators that can be used in the different risk contexts for performing the respective task. As with Table 2.1, the choice of indicators is not exhaustive and serves the purpose of illustrating the type of information needed to perform the task described in each step. The three components are normally performed sequentially but, depending on circumstances, the order may be changed. Often exposure assessments are carried out before hazards are estimated. If, for example, exposure can be prevented, it may not be necessary to perform any sophisticated hazard estimate.

Generic Challenges for Risk Assessment

Risk assessment is confronted with three major challenges that can be best described using the terms 'complexity', 'uncertainty' and 'ambiguity'. These three challenges are not related to the intrinsic characteristics of hazards or risks themselves but to the *state and quality of knowledge* available about both hazards and risks. Since risks are mental constructs, the quality of their explanatory power depends on the accuracy and validity of their (real) predictions. Compared with other scientific constructs, validating the results of risk assessments is particularly difficult because, in theory, one would need to wait indefinitely to prove that the probabilities assigned to a specific outcome were correctly assessed. If the number of predicted events is frequent and the causal chain obvious (as is the case with car accidents), validation is relatively simple

Table 2.2 *Generic components of risk assessment*

Assessment Components	Definition	Indicators
1 Hazard identification and estimation	Recognizing potential for adverse effects and assessing the strength of cause–effect relationships	– properties such as flammability – persistence – irreversibility – ubiquity – delayed effects – potency for harm – dose–response relationships
2 Exposure/ vulnerability assessment	Modelling diffusion, exposure and effects on risk targets	– exposure pathways – normalized behaviour of target – vulnerability of target
3 Risk estimation	– *Quantitative*: probability distribution of adverse effects – *Qualitative*: combination of hazard, exposure and qualitative factors (scenario construction)	– expected risk values (individual, collective) – confidence interval (%) – risk description – risk modelling as function of variations in context variables and parameters

and straightforward. If, however, the assessment focuses on risks where cause–effect relationships are difficult to discern, effects are rare and difficult to interpret, and variations in both causes and effects obscure the results, the validation of the assessment results becomes a major problem. In such instances, assessment procedures are needed to characterize the existing knowledge with respect to complexity, remaining uncertainties and ambiguities (WBGU, 2000, p195ff; Klinke and Renn, 2002).

Complexity

Complexity refers to the difficulty of identifying and quantifying causal links between a multitude of potential causal agents and specific observed effects. The nature of this difficulty may be traced back to interactive effects among these agents (synergism and antagonisms), long delay periods between cause and effect, inter-individual variation, intervening variables or other factors. Risk assessors have to make judgements about the level of complexity that they are able to process and about how to treat intervening variables (such as lifestyle, other environmental factors and psychosomatic impacts). Complexity is particularly pertinent in the phase of estimation with respect to hazards as well as risks. Examples of highly complex risk include sophisticated chemical facilities, synergistic effects of potentially toxic substances, failure risk of large interconnected infrastructures and risks of critical loads to sensitive ecosystems.

Uncertainty

Uncertainty is different from complexity but often results from an incomplete or inadequate reduction of complexity in modelling cause–effect chains. Whether the world is inherently uncertain is a philosophical question that we will not pursue here. It is essential to acknowledge in the context of risk assessment, however, that human knowledge is always incomplete and selective and thus contingent on uncertain assumptions, assertions and predictions (Functowicz and Ravetz, 1992; Laudan, 1996; Bruijn and ten Heuvelhof, 1999). It is obvious that the modelled probability distributions within a numerical relational system can only represent an approximation of the empirical relational system with which to understand and predict uncertain events (Cooke, 1991). It therefore seems prudent to include other, additional, aspects of uncertainty (Morgan and Henrion, 1990; van Asselt 2000, pp93–138; van den Sluijs et al, 2003). Although there is no consensus in the literature on the best means of disaggregating uncertainties, the following categories appear to be appropriate in distinguishing the key components of uncertainty:

- *target variability* (based on different vulnerability of targets);
- *systematic and random error in modelling* (based on extrapolations from animals to humans or from large doses to small doses, statistical inferential applications, etc);
- *indeterminacy or genuine stochastic effects* (variation of effects due to random events, in special cases congruent with statistical handling of random errors);
- *system boundaries* (uncertainties stemming from restricted models and the need for focusing on a limited number of variables and parameters); and
- *ignorance or non-knowledge* (uncertainties derived from lack of knowledge).

The first two components of uncertainty qualify as epistemic uncertainty and therefore can be reduced by improving the existing knowledge and by advancing the present modelling tools. The last three components are genuine uncertainty components of an aleatory nature and thus can be characterized to some extent using scientific approaches but cannot be further resolved. If uncertainty, in particular the aleatory components, plays a large role then the estimation of risk becomes fuzzy. The validity of the end results is then questionable and, for risk management purposes, additional information is needed such as a subjective confidence level in the risk estimates, potential alternative pathways of cause–effect relationships, ranges of reasonable estimates, loss scenarios and so forth. Examples of high uncertainty, particularly aleatory uncertainty, include many natural disasters such as earthquakes, possible health effects of mass pollutants below the threshold of statistical significance, acts of violence such as terrorism and sabotage, and long-term effects of introducing genetically modified species into the natural environment.

Ambiguity

Ambiguity (interpretative and normative) is the last term in this context. Whereas uncertainty refers to a lack of clarity over the scientific or technical basis for decision-making, ambiguity is a result of divergent or contested perspectives on the justification, severity or wider 'meanings' associated with a given threat (Stirling, 2003). The term ambiguity may be misleading here because it has different connotations in everyday English.[5] In relation to risk governance it is understood as 'giving rise to several meaningful and legitimate interpretations of accepted risk assessments results'. It can be divided into *interpretative ambiguity* (different interpretations of an identical assessment result, for example as an adverse or non-adverse effect) and *normative ambiguity* (different concepts of what can be regarded as tolerable referring, for example, to ethics, quality of life parameters, distribution of risks and benefits). A condition of ambiguity emerges where the problem lies in agreeing on the appropriate values, priorities, assumptions or boundaries to be applied to the definition of possible outcomes. What does it mean, for example, if neuronal activities in the human brain are intensified when subjects are exposed to electromagnetic radiation? Can this be interpreted as an adverse effect or is it just a bodily response without any health implication? Many scientific disputes in the fields of risk assessment and management do not refer to differences in methodology, measurements or dose–response functions, but to the question of what all of this means for human health and environmental protection. High complexity and uncertainty favour the emergence of ambiguity, but there are also quite a few simple and highly probable risks that can cause controversy and thus ambiguity. Examples for high interpretative ambiguity include low-dose radiation (ionizing and non-ionizing), low concentrations of genotoxic substances, food supplements and hormone treatment of cattle. Normative ambiguities can be associated, for example, with passive smoking, nuclear power, pre-natal genetic screening and genetically modified food.

Risk Perception

Since risk is a mental construct there is a wide variety of construction principles for conceptualizing it. Different disciplines within the natural and social sciences have formed their own concepts of risk; stakeholder groups, driven by interest and experience, have developed specific perspectives on risk; and, last but not least, representatives of civil society as well as the general public are responding to risks according to their own risk constructs and images. These images are called 'perceptions' in the psychological and social sciences and they have been intensely researched in relation to risk – as have their underlying factors (Covello, 1983; Slovic, 1987; Boholm, 1998; Rohrmann and Renn,

2000). Risk perceptions belong to the contextual aspects that risk managers need to consider when deciding whether or not a risk should be taken as well as when designing risk reduction measures.

First of all it is highly important to know that human behaviour is primarily driven by perception and not by facts or by what is understood as facts by risk analysts and scientists. Most cognitive psychologists believe that perceptions are formed by common-sense reasoning, personal experience, social communication and cultural traditions (Brehmer, 1987; Drottz-Sjöberg, 1991; Pidgeon et al, 1992; Pidgeon, 1998). In relation to risk it has been shown that humans link certain expectations, ideas, hopes, fears and emotions with activities or events that have uncertain consequences. People do not, however, use completely irrational strategies to assess information: most of the time they follow relatively consistent patterns of creating images of risks and evaluating them. These patterns are related to certain evolutionary bases of coping with dangerous situations. Faced with an imminent threat, humans react with four basic strategies: *flight, fight, play dead* and, if appropriate, *experimentation* (on the basis of trial and error).

In the course of cultural evolution the basic patterns of perception were increasingly enriched with cultural patterns. These cultural patterns can be described by so-called *qualitative evaluation characteristics* (Slovic, 1992). They describe properties of risks or risky situations going beyond the two classical factors of risk assessment on the basis of which risk is usually judged (level of probability and degree of possible harm). Here, psychologists differentiate between two classes of qualitative perception patterns: on the one hand *risk-related* patterns, which are based on the properties of the source of risk; on the other *situation-related* patterns, based on the idiosyncrasies of the risky situation (Fischhoff et al, 1978; Slovic, 1987; Slovic, 1992).

One example of a risk-related pattern is the perceived 'dread' of the consequences of a possible harmful event. If, for example, a person is riding in a car and thinking about possible accidents, he or she will always be under the impression that he or she would, with high probability, get away unscathed in a car accident (the 'fender-bender mentality'). However, if the same person is sitting in an aeroplane he or she will be under the impression that if something happens here there is no getting away. This feeling of apprehensiveness does not subside even when this person knows the odds and is convinced that statistically many more people die in car accidents than in aeroplane crashes. Situation-related patterns of perception include aspects such as voluntariness and the ability to exercise self-control. If a person is of the opinion that he or she can control the risk, then he or she will perceive it as less serious. This mode of thinking frequently takes effect where eating habits are concerned. People believe they could easily do without sweets, alcohol or other food considered unhealthy if only they wanted to, while on the other hand mostly harmless chemical food additives are perceived as a threat to one's health. With respect to collective risks people show special concern for risks that they believe are not adequately controlled by public authorities (as in the case of GMOs).

Considered together these qualitative evaluation characteristics can be subdivided into a limited number of consistent risk perception classes. In the literature they are also called 'semantic risk patterns'. The following patterns have been examined particularly thoroughly:

- immediate threats, such as risk associated with nuclear energy or large dams;
- risks dealt with as a blow of fate, such as natural disasters;
- risks presenting a challenge to one's own strength, such as sports activities;
- risk as a gamble, such as lotteries, stock exchanges or insurance; and
- risks as an early indication of insidious danger, such as food additives, ionizing radiation or viruses. (Renn, 2004a)

These patterns have functions similar to drawers in a filing cabinet. When faced with a new risk or when obtaining new information about a risk, most people try to file this new information into one of the existing drawers.[6] In addition to the cognitive processing of risk characteristics and risk situations, studies have shown that people tend to *stigmatize* risk sources that are associated with specific dreadful associations (Kunreuther and Heal, 2003). A salient example of stigma is the reaction to products that are deemed to be carcinogenic, although there is often limited, if any, scientific evidence to support this position. The mere suspicion that a substance could cause cancer is often sufficient for generating fear and asking for strict regulatory actions. Stigmatization leads to a cycle of public outrage and regulatory responses, feeding into a process that has been described as social amplification of risk (Kasperson et al, 1988 and 2003). Stimulated by media reporting, the public's perception of the risk is often amplified in ways that are difficult to explain if one were focusing on the standard elements of any technical risk assessment – probability and direct losses.

The problems associated with risk perception are compounded because of the difficulty individuals have in interpreting low probabilities when making their decisions (Kunreuther et al, 2001). In fact, there is evidence that people may not even want data on the likelihood of an event occurring. If people do not think probabilistically, then how do they make their choices? Psychological research has revealed the following patterns of drawing inferences about probabilities and risks:

- the easier and faster a risk is recognized, the more conscious individuals are of it and the greater is the chance of its probability being overestimated. If, for example, an individual has known someone who died after being struck by lightning, that individual will perceive the risk of being struck by light-ning as being particularly large (*availability bias*).

- the more a risk provokes associations with known events, the more likely its probability will be overestimated. This is why, for example, the use of the term 'incinerating' in waste-disposal facilities readily evokes an associ-ation with harmful chemicals, especially dioxins and furans, even if there is no way that they could be released into the environment by the facilities concerned (*anchoring effect*).

- the more constant and similar the losses from risk sources, the more likely the impact of average losses will be underestimated. While road traffic accidents are not deemed acceptable, they are more or less passively accepted. If the average annual number of road deaths in a given country were to occur at one point in time instead of being spread out over the year, then a considerably greater level of rejection could be expected. Thus, people are not indifferent as regards the *distribution of risks over time*: they prefer even loss distribution over individual disasters (Kahneman and Tversky, 1979).

- the greater the uncertainty of loss expectation, the more likely the average loss assessment will be in the region of the median of all known loss expec-tations. In this way, loss expectations in objectively low risks are often overestimated while objectively high risks are often underestimated (*assess-ment bias*) (Tversky and Kahneman, 1974; Ross, 1977; Kahneman and Tversky, 1979; Renn, 2004a).

While important for actually evaluating and managing a risk, overestimation or underestimation of loss expectations is not, however, the most important aspect of risk perception. Instead the context-dependent nature of risk assess-ment is the deciding factor. This context includes the qualitative risk evaluation characteristics, the semantic images and the stigma effects. More recently, psychologists have also discovered that affect and emotions play an important role in people's decision processes (Slovic et al, 2002; Loewenstein et al, 2001). These factors are particularly relevant when individuals face a decision that involves a difficult trade-off between attributes or where there is interpretative ambiguity as to what constitutes a 'right' answer. In these cases, people often appear to resolve problems by focusing on those cues that send the strongest affective signals (Hsee and Kunreuther, 2000).

The most important policy question is how to treat risk perceptions in a policy arena that includes responses of different actors and the general public (Slovic et al, 1982; Fischhoff, 1985 and 1995). There are two suggestions, from opposite ends of a spectrum. The first position states that the scientific concepts of risk are the only ones that can claim inter-subjective validity and applicability and, therefore, requires risk managers to obtain an assurance that erroneous risk perceptions are corrected via risk communication and education (Cross, 1998; Coglianese, 1999). The second position states that there is no overarching universally applicable quality criterion available in order to evaluate the appropriateness or validity of risk concepts. As a result, scientific concepts (often called *narratives* in this school of thought) should compete with the concepts of stakeholders and public groups (Liberatore and Funtowicz, 2003). If collective decisions on risk are necessary, the concept that is used to make these decisions should be negotiated among all relevant concept holders. None of these groups, including the science communities, is allowed to claim any privileged position in this negotiation.

This position has major impacts on risk policy-making and communication. Policy-making needs, inter alia, to organize systematic feedback from society and, equally, to include risk perceptions as an important input to deciding on whether something should be done about a certain risk and, if so, what (Jaeger et al, 2001). How this can be accomplished is explained in the next section, on risk appraisal. Risk communication is also affected in two ways: first, it is bound to elicit – and enable the exchange of – concerns and conceptual aspects of risk among and between all relevant actors; second, risk managers are well advised to ensure that the best available knowledge is widely distributed to those who raise such concerns.

Risk Appraisal

The term risk appraisal has sometimes been used in the risk governance literature to include all knowledge elements necessary for risk characterization and evaluation as well as risk management (Stirling, 1998 and 2003). For society to make prudent choices about risks, however, it is not enough to consider only the results of scientific risk assessment. In order to understand the concerns of the various stakeholders and public groups, information about both risk perceptions and the further implications of the direct consequences of a risk – including its social mobilization potential (in other words how likely it is that the activity will give rise to social opposition or protest) – is needed and should be collected by risk management agents. In addition, other aspects of the risk-causing activity that seem to be relevant for characterizing and evaluating the risk and selecting risk reduction options should be pulled together and fed into the analysis. Based on such a wide range of information, risk managers can make more informed judgements and design the appropriate risk management options (Clark, 2001).

Risk appraisal thus includes the scientific assessment of the risks to human health and the environment and an assessment of related concerns, as well as social and economic implications. The appraisal process is and should be clearly dominated by scientific analyses, but, in contrast to the traditional risk governance model, the scientific process includes both the natural/technical as well as the social sciences, including economics. We envision risk appraisal as having two process stages: first, natural and technical scientists use their skills to produce the best estimate of the physical harm that a risk source may induce (as described in the section on risk assessment, page 24); second, social scientists and economists identify and analyse the issues that individuals or society as a whole link with a certain risk. For this purpose the repertoire of the social sciences, such as survey methods, focus groups, econometric analysis, macroeconomic modelling or structured hearings with stakeholders, may be used.

Based on the results of risk assessment and the identification of individual and social concerns this second process stage also investigates and calculates the *social and economic implications* of risks. Of particular interest in this context are financial and legal implications, in other words economic losses and liabilities, as well as social responses such as political mobilization. These secondary implications have been addressed by the concept of *social amplification of risk* (Kasperson et al, 2001 and 2003). This concept is based on the hypothesis that events pertaining to hazards interact with psychological, social, institutional and cultural processes in ways that can heighten or attenuate individual and social perceptions of risk and shape risk behaviour. Behavioural patterns, in turn, generate secondary social or economic consequences that extend far beyond direct harm to human health or the environment, including significant indirect impacts such as liability, insurance costs, loss of confidence in institutions or alienation from community affairs (Burns et al, 1993). Such amplified secondary effects can then trigger demands for additional institutional responses and protective actions, or, conversely (in the case of risk attenuation), place impediments in the path of needed protective actions. Secondary impacts, whether amplified or not, are of major concern to those who are obliged to take over the costs or cope with the consequences of being accountable.

Risk appraisal intends to produce the best possible scientific estimate of the physical, economic and social consequences of a risk source. It should not be confused with direct stakeholder involvement, which will be covered later. Involvement by stakeholders and the population is only desirable at this stage if knowledge from these sources is needed to improve the quality of the assessments.

In a recent draft document published by the UK Treasury Department (UK Treasury Department, 2004) the authors recommend a risk appraisal procedure that includes the results of risk assessment, the direct input from data on public perception and the assessment of social concerns. The document offers a tool for evaluating public concerns against six factors which are centred around the hazard or hazards leading to a risk, the risk's effects and its management:[7]

1   perception of familiarity and experience with the hazard;
2   understanding the nature of the hazard and its potential impacts;

3　repercussions of the risk's effects on equity (inter-generational, intra-generational, social);
4　perception of fear and dread in relation to a risk's effect;
5　perception of personal or institutional control over the management of a risk; and
6　degree of trust in risk management organizations.

A similar list of appraisal indicators was suggested by a group of Dutch researchers and the Dutch Environmental Protection Agency (van den Sluijs et al, 2003 and 2004). In the late 1990s the German Council for Global Environmental Change (WBGU) also addressed the issue of risk appraisal and developed a set of eight criteria to characterize risks beyond the established assessment criteria:

1　*extent of damage*: Adverse effects in natural units (death, injury, production loss, etc);
2　*probability of occurrence*: Estimate of relative frequency, which can be discrete or continuous;
3　*incertitude*: How do we take account of uncertainty in knowledge, in modelling of complex systems or in predictability in assessing a risk?
4　*ubiquity*: Geographical dispersion of damage;
5　*persistence*: How long will the damage last?
6　*reversibility*: Can the damage be reversed?
7　*delay effects*: Latency between initial event and actual damage; and
8　*potential for mobilization*: The broad social impact. Will the risk generate social conflict or outrage? (WBGU, 2000)

After the WBGU proposal had been reviewed and discussed by many experts and risk managers, it was suggested to unfold the compact 'mobilization index' and divide it into four major elements:

1　*inequity and injustice* associated with the distribution of risks and benefits over time, space and social status;
2　*psychological stress and discomfort* associated with the risk or the risk source (as measured by psychometric scales);
3　*potential for social conflict and mobilization* (degree of political or public pressure on risk regulatory agencies); and
4　*spill-over effects* that are likely to be expected when highly symbolic losses have repercussions on other fields such as financial markets or loss of credibility in management institutions.

These four sub-criteria reflect many factors that have been proven to influence risk perception.[8]

When dealing with complex, uncertain and/or ambiguous risks it is essential to complement data on physical consequences with data on secondary impacts, including social responses to risk, and insights into risk perception. The suggestions listed above can provide some orientation for the criteria to be considered. Depending on the risk under investigation, however, additional criteria can be included or proposed criteria ignored.

Characterizing and Evaluating Risks

The most controversial part of handling risks is the process of delineating and justifying a judgement about the tolerability or acceptability of a given risk (HSE, 2001). The term 'tolerable' refers to an activity that is seen as worth pursuing (for the benefit it carries) yet requires additional efforts for risk reduction within reasonable limits. The term 'acceptable' refers to an activity where the remaining risks are so low that additional efforts for risk reduction are not seen as necessary. For purely natural hazards the two terms appear at first glance to be meaningless, since humans have no choice in tolerating or accepting these risks. Human activities, however, do influence the impact of natural hazards through changes in vulnerability and exposure options (such as building codes or zoning laws). Looking into the resulting risks as a function of vulnerabilities, a judgement on tolerability and acceptability with respect to the selection of protective measures becomes meaningful again. The distinction between tolerability and acceptability can thus be applied to a large array of risk sources. If tolerability and acceptability are set out in a risk diagram (with probabilities on the *y*-axis and extent of consequences on the *x*-axis), the well-known traffic light model emerges (Figure 2.1).⁹ In this variant of the model the red zone (light grey in our black and white representation here) signifies intolerable risk, the amber zone (medium grey shading in our diagram) indicates tolerable risk in need of further management actions (in accordance with the ALARP principle) and the green zone (darkest grey) shows acceptable or even negligible risk.

To draw the line between 'intolerable' and 'tolerable' as well as 'tolerable' and 'acceptable' is one of the most difficult tasks of risk governance. The UK Health and Safety Executive has developed a procedure for chemical risks based on risk–risk comparisons (Löfstedt, 1997). Some Swiss cantons such as Basle County have experimented with round tables as a means to reach consensus on drawing the two lines, whereby participants in the round table represented industry, administrators, county officials, environmentalists and neighbourhood groups (RISKO, 2000). Irrespective of the selected means to support this task, the judgement on acceptability or tolerability is contingent on making use of a variety of different knowledge sources. And it is necessary to include the risk estimates derived from the risk assessment stage and additional assessment data from the concern assessment within the appraisal stage.

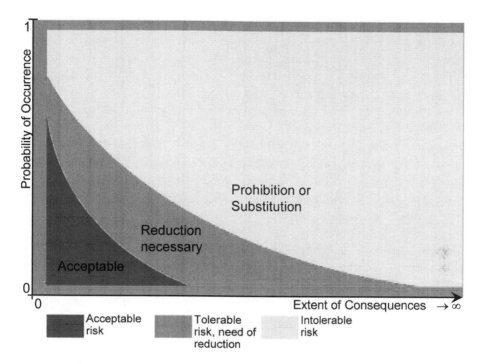

Figure 2.1 Acceptable, tolerable and intolerable risks (traffic light model)

Existing taxonomies of risk differ considerably in where they position the decision-making with regard to what is acceptable and what is tolerable within the overall risk process. Some assign it to the risk assessment part, others to the risk management part and others place it at the level of policy and option assessment, reaching far beyond the narrow risk acceptance criteria. For the generic approach to risk handling that this chapter pursues, the question of appropriate placement should be handled in a flexible manner.

Why? As with the framing part, judgements on acceptability rely on two major inputs: *values* and *evidence*. What society is supposed to tolerate or accept can never be derived from looking at the evidence alone, while evidence is essential if we are to know whether a value has been violated or not (or to what degree). With respect to values and evidence we can distinguish three cases: (i) ambiguity on evidence but not on values (interpretative ambiguity); (ii) ambiguity on values but not on evidence (normative ambiguity); and (iii) ambiguities on values and evidence.

Case I: Interpretative ambiguity

In cases where there is unanimous agreement about the underlying values and even the threshold of what is regarded as tolerable or acceptable, evidence in the form of risk estimates may be sufficient to locate the risk within the traffic light diagram. A judgement can then best be made by those who have most expertise in risk and concern assessments, in which case it makes sense to place this task within the domain of risk appraisal. The judgement will thus be based on best scientific modelling of epistemic uncertainties and the best qualitative characterization of aleatory uncertainties. Characterization also includes an analysis of the concerns associated with different outcomes and the likely secondary implications. It will be helpful for risk managers to receive best expert advice on potentially effective risk reduction measures and other management options that may lead to satisfactory results. It is, however, not the task of the risk appraisal team to make a selection of options, let alone decide on which option should be implemented.

Leaving the resolution of interpretative ambiguity to the risk and concern assessors presents a major challenge to the science-based assessment process. It may be extremely difficult for experts to find an agreement on interpreting ambiguous results. It is not uncommon for the public to hear Expert 1 say that there is 'nothing to worry about regarding a particular risk' while learning from Expert 2 that 'this risk should be on your radar screen'. One way to capture these discrepancies in risk interpretations is to construct an *exceedance probability (EP) curve* (Grossi and Kunreuther, 2005). An EP curve specifies the probabilities that a certain level of losses will be exceeded. The losses can be measured in terms of dollars of damage, fatalities, illness or some other unit of analysis.

To illustrate with a specific example, suppose one was interested in constructing an EP curve for dollar losses to homes in Seattle from an earthquake. Using probabilistic risk assessment, one combines the set of events that could produce a given dollar loss and then determines the resulting probabilities of exceeding losses of different magnitudes. Based on these estimates, one can construct the mean EP depicted in Figure 2.2. By its nature, the EP curve inherently incorporates uncertainty associated with the probability of an event occurring and the magnitude of dollar losses. This uncertainty is reflected in the 5 per cent and 95 per cent confidence interval curves in the figure.

The EP curve also serves as an important tool for evaluating risk management options, thus assisting managers to optimize risk reduction. It puts pressure on experts to state the assumptions on which they are basing their estimates of the likelihood of certain events occurring and the resulting consequences. In fact, EP curves, such as those depicted in Figure 2.2, supplemented by a discussion of the nature of these assumptions, should enable the assessors to both characterize interpretative ambiguities and provide a framework for risk managers to test the efficiency of risk reduction options.

Case 2: Normative ambiguity

If the underlying values of what could be interpreted as tolerable or acceptable are disputed, while the evidence of what is at stake is clearly given and non-

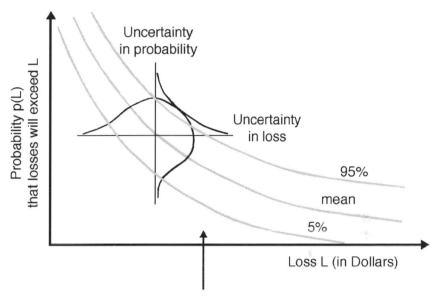

Figure 2.2 Example of loss exceedance probability curves

controversial, the judgement needs to be based on a discourse about values and their implications. Such a discourse falls clearly in the domain of risk management. A good example may be the normative implications of risks related to smoking. Science is very familiar with these risks and there is little uncertainty and interpretative ambiguity about dose–effect relationships. Yet there is considerable debate whether smoking is tolerable or not. Being a voluntary activity some countries leave it to the decision of each consumer while others initiate major activities to reduce and even ban smoking. Another example is that of bicycle helmets. The statistical data on this subject is relatively straightforward: there are no major uncertainties or interpretative ambiguities. Yet many countries do not want to impinge on the freedom of each cyclist to personally decide whether or not to wear a helmet, while other countries pursue a more paternalistic policy.

Case 3: Interpretative and normative ambiguity

A third case arises where both the evidence and the values are disputed. This would imply that assessors should engage in an activity to find some common ground for characterizing and qualifying the evidence and risk managers need to establish agreement about the appropriate values and their application. A good example for this third case may be the interpretative and normative implications of global climate change. An international expert group such as the Intergovernmental Panel on Climate Change (IPCC) has gone to considerable effort to articulate a common characterization of climatic risks and their uncertainties. Given the remaining uncertainties and the complexities of the causal relationships between greenhouse gases and climate change, it is then a question of values whether governments place their priorities on prevention or on mitigation (Keeney and McDaniels, 2001).

Since the third of the above cases includes both of the other two, the process of judging the tolerability and acceptability of a risk can be structured into two distinct components: *risk characterization* and *risk evaluation*. The first step, risk characterization, determines the evidence-based component for making the necessary judgement on the tolerability and/or acceptability of a risk; the risk evaluation step determines the value-based component for making this judgement. Risk characterization includes tasks such as point estimates of risks, descriptions of remaining uncertainties (as undertaken for instance in climate change models or risk studies on endocrine disruptors) and potential outcome scenarios including the social and economic implications, suggestions for safety factors to include inter-target variation, assurance of compatibility with legal prescriptions, risk–risk comparisons, risk–risk trade-offs, identification of discrepancies between risk assessment and risk perceptions as well as of potential equity violations, and suggestions for reasonable standards to meet legal requirements (Stern and Fineberg, 1996). The evidence collected and summarized here goes beyond the classic natural science reservoir of knowledge and includes economic and social science expertise. This is also the reason why in the process of risk characterization an interdisciplinary team of scientists is needed to draw a complete picture of what is known and what is and may remain unknown. In the course of risk characterization, scientists are asked to design a multi-criteria profile of the risk in question, make a judgement about the seriousness of the risk and suggest potential options to deal with the risk.

The second step, risk evaluation, broadens the picture to include pre-risk aspects such as choice of technology, social need for the specific risk agent (substitution possible?), risk–benefit balances, political priorities, potential for conflict resolution and social mobilization potential. The main objective here is to arrive at a judgement on tolerability and acceptability based on balancing pros and cons, testing potential impacts on quality of life, discussing different development options for the economy and society, and weighing the competing arguments and evidence claims in a balanced manner. It should be noted that this elaborate procedure is only necessary if tolerability and/or acceptability is disputed and if society faces major dissents and conflicts among important stakeholders. If so, the direct involvement of stakeholders and the public will be a prerequisite for successful risk governance.

The separation of evidence and values underlying the distinction between characterization and evaluation is, of course, functional and not necessarily organizational. Since risk characterization and evaluation are closely linked and each depends on the other, it may even be wise to perform these two steps simultaneously in a joint effort by both assessors and risk managers. As some analysts have pointed out (Löfstedt and Vogel, 2001; Vogel, 2003), the US regulatory system tends to favour an organizational combination of characterization and evaluation, while European risk managers tend to maintain the organizational separation (particularly in the area of food).

The distinction between the three challenges of risk assessment – complexity, uncertainty and ambiguity – can also assist assessors and managers in assigning, or dividing, the judgement task. If a given risk is characterized by high complexity, low remaining uncertainties and hardly any ambiguities (except for interpretative differences over an established scientific risk assessment result), it is wise to let the assessment team dominate the process of making tolerability/acceptability judgements. If, in contrast, the risk is characterized by major unresolved uncertainties and if the results lead to highly diverse interpretations of what they mean for society, it is advisable to let risk managers take the lead.

Table 2.3 summarizes these two steps which, in conclusion, are closely interrelated and may be merged if the circumstances require it. The list of indicators again represents only a small selection of potential dimensions and is displayed here for illustrative purposes.

Risk Management

Risk management starts with a review of all relevant information, in particular that from the combined risk appraisal, consisting of both a risk assessment and a concern assessment, the latter based on risk perception studies, economic impact assessments and the scientific characterization of social responses to the risk source. This information, together with the judgements made in the phase of risk characterization and evaluation, form the input material on which risk management options are assessed, evaluated and selected. At the outset, risk management is presented with three potential outcomes:

1   *intolerable situation*: this means that either the risk source (such as a technology or a chemical) needs to be abandoned or replaced or, in cases where that is not possible (for example natural hazards), vulnerabilities need to be reduced and exposure restricted;
2   *tolerable situation*: this means that the risks need to be reduced or handled in some other way within the limits of reasonable resource investments (ALARP, including best practice). This can be done by private actors (such as corporate risk managers) or public actors (such as regulatory agencies) or both (public–private partnerships); or

Table 2.3 *Tolerability/acceptability judgement*

Assessment Components	Definition	Indicators
1 Risk characterization	Collecting and summarizing all relevant evidence necessary for making an informed choice on tolerability or acceptability of the risk in question and suggesting potential options for dealing with the risk from a scientific perspective	
	a risk profile	– risk estimates – confidence intervals – uncertainty measures – hazard characteristics – range of 'legitimate' interpretations – risk perceptions – social and economic implications
	b judging the seriousness of risk	– compatibility with legal requirements – risk–risk trade-offs – effects on equity – public acceptance
	c conclusions and risk reduction options	suggestions for: – tolerable risk levels – acceptable risk levels – options for handling risks
2 Risk evaluation	Applying societal values and norms to the judgement on tolerability and acceptability and, consequently, determining the need for risk reduction measures	– choice of technology – potential for substitution – risk–benefit comparison – political priorities – compensation potential – conflict management – potential for social mobilization

3   *acceptable situation*: this means that the risks are so small – perhaps even regarded as negligible – that any risk reduction effort is unnecessary. However, risk sharing via insurances and/or further risk reduction on a voluntary basis present options for action which can be worthwhile pursuing even in the case of an acceptable risk.

With regard to these outcomes risk managers may face either a situation of unanimity, in other words all relevant actors agree on how a given risk situation should be qualified, or a situation of conflict in which major actors challenge the classification undertaken by others. The degree of controversy is one of the drivers for selecting the appropriate instruments for risk prevention or risk reduction.

For a systematic analysis of the risk management process it is advisable to focus on tolerable risks and those where tolerability is disputed, since the other cases are fairly easy to deal with. In the case of intolerable risks – and often in the case of tolerable but highly disputed risks – risk managers should opt for prevention strategies as a means to replace the hazardous activity with another activity leading to identical or similar benefits. One should first make sure, however, that the replacement does not introduce more risks or more uncertainties than the agent that it replaces (Wiener, 1998). In the case of acceptable risks it should be left to private actors to initiate additional risk reduction or to seek insurance for covering potential but acceptable losses (although this does not eliminate the need for all concerned to have sufficient information and resources to do so). If risks are classified as tolerable, or if there is dispute as to whether they are tolerable or acceptable, risk management needs to design and implement actions that make these risks acceptable over time. Should this not be feasible then risk management, aided by communication, needs at least to credibly convey the message that major effort is being undertaken to bring these risks closer to being acceptable. This task can be described in terms of classic decision theory, in other words in the following steps (Morgan, 1990; Keeney, 1992; Hammond et al, 1999):

Identification and generation of risk management options
Generic risk management options include risk avoidance, risk reduction, risk transfer and – also an option to take into account – self retention. Whereas to avoid a risk means either selecting a path which does not touch on the risk (for example by abandoning the development of a specific technology) or taking action in order to fully eliminate a certain risk, risk transfer deals with ways of passing the risk on to a third party. Self retention as a management option essentially means taking an informed decision to do nothing about the risk and to take full responsibility both for the decision and any consequences occurring thereafter. Risk management by means of risk reduction can be accomplished by many different means. Among these are:

- technical standards and limits that prescribe the permissible threshold of concentrations, emissions, take-up or other measures of exposure;
- performance standards for technological and chemical processes, such as minimum temperatures in waste incinerators;
- technical prescriptions referring to the blockage of exposure (for example via protective clothing) or the improvement of resilience (for example via immunization or earthquake-tolerant constructions);
- governmental economic incentives including taxation, duties, subsidies and certification schemes;
- third-party incentives (private monetary or in kind incentives);
- compensation schemes (monetary or in kind);

- insurance and liability; and
- cooperative and informative options ranging from voluntary agreements to labelling and education programmes.

All these options can be used individually or in combination to accomplish even more effective risk reduction. Options for risk reduction can be initiated by private and/or public actors.

Assessment of risk management options with respect to predefined criteria

Each of the options will have desired and unintended consequences which relate to the risks that they are supposed to reduce. In most instances, an assessment should be done according to the following criteria:

- *Effectiveness*: Does the option achieve the desired effect?
- *Efficiency*: Does the option achieve the desired effect with the least resource consumption?
- *Minimization of external side effects*: Does the option infringe on other valuable goods, benefits or services such as competitiveness, public health, environmental quality, social cohesion, etc? Does it impair the efficiency and acceptance of the governance system itself?
- *Sustainability*: Does the option contribute to the overall goal of sustainability? Does it assist in sustaining vital ecological functions, economic prosperity and social cohesion?
- *Fairness*: Does the option burden the subjects of regulation in a fair and equitable manner?
- *Political and legal implementability*: Is the option compatible with legal requirements and political programmes?
- *Ethical acceptability*: Is the option morally acceptable?
- *Public acceptance*: Will the option be accepted by those individuals who are affected by it? Are there cultural preferences or symbolic connotations that have a strong influence on how the risks are perceived?

Measuring management options against these criteria may create conflicting messages and results. Many measures that prove to be effective may turn out to be inefficient or unfair to those who will be burdened. Other measures may be sustainable but not accepted by the public or important stakeholders. These problems are aggravated when dealing with global risks. What appears to be efficient in one country may not work at all in another country. Risk managers are therefore well advised to make use of the many excellent guidance documents on how to handle risk trade-offs and how to employ decision analytic tools for dealing with conflicting evidence and values (Viscusi, 1994; Wiener, 1998; van der Sluijs et al, 2003; Goodwin and Wright, 2004).

Evaluation of risk management options

Similar to risk evaluation, this step integrates the evidence on how the options perform with regard to the evaluation criteria with a value judgement about the relative weight each criterion should be assigned. Ideally, the evidence should come from experts and the relative weights from politically legitimate decision-makers. In practical risk management, the evaluation of options is done in close cooperation between experts and decision-makers. As pointed out later, this is the step in which direct stakeholder involvement and public participation is particularly important and is therefore best assured by making use of a variety methods (Rowe and Frewer, 2000; OECD, 2002).

Selection of risk management options

Once the different options are evaluated, a decision has to be made as to which options are selected and which rejected. This decision is obvious if one or more options turn out to be dominant (better on all criteria). Otherwise, trade-offs have to be made that need legitimization (Graham and Wiener, 1995). A legitimate decision can be made on the basis of formal balancing tools (such as cost–benefit or multi-criteria-decision analysis), by the respective decision-makers (given their decision is informed by a holistic view of the problem) or in conjunction with participatory procedures.

Implementation of risk management options

It is the task of risk management to oversee and control the implementation process. In many instances implementation is delegated, as when governments take decisions but leave their implementation to other public or private bodies or to the general public. However, the risk management team has at any rate the implicit mandate to supervise the implementation process or at least monitor its outcome.

Monitoring of option performance

The last step refers to the systematic observation of the effects of the options once they are implemented. The monitoring system should be designed to assess intended as well as unintended consequences. Often a formal policy assessment study is issued in order to explore the consequences of a given set of risk management measures on different dimensions of what people value. In addition to generating feedback for the effectiveness of the options taken to reduce the risks, the monitoring phase should also provide new information on early-warning signals for both new risks and old risks viewed from a new perspective. It is advisable to have the institutions performing the risk and concern assessments participate in monitoring and supervision so that their analytic skills and experience can be utilized in evaluating the performance of the selected management options.

These steps follow a logical sequence but can be arranged in different orders depending on both situation and circumstance. It might be helpful to

visualize the steps not as a linear progression but as a circle forming an iterative process in which reassessment phases are intertwined, with new options
emerging, new crisis situations arising or new demands being placed on risk
managers. Similarly, sometimes the assessment of different options causes the
need for new options to be created in order to achieve the desired results. In
other cases, the monitoring of existing rules impacts on the decision to add
new criteria to the portfolio. Rarely do issues for risk appraisal and management thus follow the exact sequence used here for the description of the
process. Option generation, information processing and options selection
should indeed be seen as a dynamic process with many iterative loops.

Table 2.4 *Generic components of risk management*

Management Components	Definition	Indicators
1 Option generation	Identification of potential risk handling options, in particular risk reduction, i.e. prevention, adaptation and mitigation, as well as risk avoidance, transfer and retention	– standards – performance rules – restrictions on exposure or vulnerability – economic incentives – compensation – insurance and liability – voluntary agreements – labels – information/education
2 Option assessment	Investigations of impacts of each option (economic, technical, social, political, cultural)	– effectiveness – efficiency – minimization of side effects – sustainability – fairness – legal and political implementability – ethical acceptability – public acceptance
3 Option evaluation and selection	Evaluation of options (multi-criteria analysis)	– assignment of trade-offs – incorporation of stakeholders and the public
4 Option implementation	Realization of the most preferred option	– accountability – consistency – effectiveness
5 Monitoring and feedback	– observation of effects of implementation (link to early warning) – ex-post evaluation	– intended impacts – non-intended impacts – policy impacts

Table 2.4 summarizes the steps of risk management in accordance with the basic model used by decision theory. The list of indicators represents the most frequently used heuristic rules for selecting input and for measuring performance.

Risk Management Strategies

Based on the distinction between complexity, uncertainty and ambiguity it is possible to design generic strategies of risk management to be applied to classes of risks, thus simplifying the risk management process as outlined above. Four such classes can be distinguished:

1 *Simple risk problems*. This class of risk problems requires hardly any deviation from traditional decision-making. Data is provided by statistical analysis, goals are determined by law or statutory requirements, and the role of risk management is to ensure that all risk reduction measures are implemented and enforced. Traditional risk–risk comparisons (or risk–risk trade-offs), risk–benefit analysis and cost-effectiveness studies are the instruments of choice for finding the most appropriate risk reduction measures. Additionally, risk managers can rely on best practice and, in cases of low impact, on trial and error. It should be noted, however, that simple risks should not be equated with small or negligible risks. The major issues here are that the potential negative consequences are obvious, the values that are applied are non-controversial and the remaining uncertainties low. Examples are car accidents, known food and health risks, regularly reoccurring natural disasters or safety devices for high buildings.

2 *Complex risk problems*. For this risk class major input for risk management is provided by the scientific characterization of the risk. Complex risk problems are often associated with major scientific dissent about complex dose–effect relationships or the alleged effectiveness of measures to decrease vulnerabilities (complexity refers to both the risk agent and its causal connections and the risk-absorbing system and its vulnerabilities). The objective for resolving complexity is to receive a complete and balanced set of risk and concern assessment results that fall within the legitimate range of plural truth claims.

In a situation where there are no complete data the major challenge is to define the factual base for making risk management or risk regulatory decisions. So the main emphasis is on improving the reliability and validity of the results that are produced in the risk appraisal phase. Risk and concern assessors as well as managers need to make sure that all relevant knowledge claims are selected, processed and evaluated. They may not get a single answer but they might be able to get a better overview on the issues of scientific controversy. If these efforts lead to an acknowledgment of wide margins of uncertainty, the management tools of the uncertainty strategy should be applied. If input variables to decision-making can be properly defined and affirmed, risk characterization and evaluation can be done on the basis of risk–benefit balancing and normative standard setting (*risk-based/risk-informed regulation*). Traditional methods such as risk–risk-comparison, cost-effectiveness and cost–benefit analysis are also well-suited to facilitate the overall judgement for placing the risk in the traffic-light model (acceptable, tolerable or intolerable). These instruments, if properly used, provide effective, efficient and fair solutions with respect to finding the best trade-off between opportunities and risks. The choice of instruments includes all the classic options outlined in the section on risk management.

It is, however, prudent to distinguish management strategies for *handling the risk agent* (such as a chemical or a technology) from those *needed for the risk-absorbing system* (such as a building, an organism or an ecosystem). Addressing complex structures of risk agents requires methods for improving causal modelling and data quality control. With respect to risk-absorbing systems the emphasis is on the improvement of *robustness* in responding to whatever the target is going to be exposed to. Measures to improve robustness include inserting conservatisms or safety factors as an assurance against individual variation (normally a factor of 10–100 for occupational risk exposure and 100–1000 for public risk exposure), introducing redundant and diverse safety devices to improve structures against multiple stress situations, reducing the susceptibility of the target organism (for example iodine tablets for radiation protection), establishing building codes and zoning laws to protect against natural hazards, and improving the organizational capability to initiate, enforce, monitor and revise management actions (high reliability, learning organizations).

3 *Risk problems due to high unresolved uncertainty*. If there is a high degree of remaining uncertainties, risk management needs to incorporate hazard criteria (which are comparatively easy to determine), including aspects such as reversibility, persistence and ubiquity, and select management options empowering society to deal even with worst-case scenarios (such as containment of hazardous activities, close monitoring of risk-bearing activities or securing reversibility of decisions in case risks turn out to be higher than expected). We would argue that the management of risks characterized by multiple and high uncertainties should be guided by the *precautionary approach*. Since high unresolved uncertainty implies that the true dimensions of the risks are not yet known, one should pursue a cautious strategy that allows learning by restricted errors. The main management philosophy for this risk class is to allow small steps in implementation (the containment

approach) that enable risk managers to stop or even reverse the process as new knowledge is produced or the negative side effects become visible. The primary thrust of precaution is to avoid irreversibility (Klinke and Renn, 2002).[11]

With respect to risk-absorbing systems, the main objective is to make these systems resilient so they can withstand or even tolerate surprises. In contrast to robustness, where potential threats are known in advance and the absorbing system needs to be prepared to face these threats, *resilience* is a protective strategy against unknown or highly uncertain hazards. Instruments for resilience include the strengthening of the immune system, diversification of the means for approaching identical or similar ends, reduction of the overall catastrophic potential or vulnerability even in the absence of a concrete threat, design of systems with flexible response options, and the improvement of conditions for emergency management and system adaptation. Robustness and resilience are closely linked but they are not identical and require partially different types of actions and instruments.[10]

4 *Risk problems due to interpretative and normative ambiguity.* If risk information is interpreted differently by different stakeholders in society – in other words there are different viewpoints about the relevance, meaning and implications of factual explanations and predictions for deciding about the tolerability of a risk as well as management actions – and if the values and priorities of what should be protected or reduced are subject to intense controversy, risk management needs to address the causes for these conflicting views (von Winterfeldt and Edwards, 1984).

Genetically modified organisms for agricultural purposes may serve as an example to illustrate the intricacies related to ambiguity. Surveys on the subject demonstrate that people associate high risks with the application of gene technology for social and moral reasons (Hampel and Renn, 2000). Whether the benefits to the economy balance the costs to society in terms of increased health risks was not mentioned as a major concern of the polled public. Instead, people disagreed about the social need for genetically modified food in Western economies, where abundance of conventional food is prevalent. They were worried about the loss of personal capacity to act when selecting and preparing food, about the long-term impacts of industrialized agriculture and the moral implications of tampering with nature (Sjöberg, 1999). These concerns cannot be addressed by either scientific risk assessments or by determining the right balance between over- and under-protection. The risk issues in this debate focus on the differences between visions of the future, basic values and convictions, and the degree of confidence in the human ability to control and direct its own technological destiny. These wider concerns require the inclusion within the risk management process of those who express or represent them.

Table 2.5 *Risk characteristics and their implications for risk management*

Knowledge Characterization	Management Strategy	Appropriate Instruments	Stakeholder Participation
1 'Simple' risk problems	*Routine-based:* (tolerability/ acceptability judgement)	• Applying 'traditional' decision-making – Risk–benefit analysis – Risk–risk trade-offs	Instrumental discourse
	(risk reduction)	– Trial and error – Technical standards – Economic incentives – Education, labelling, information – Voluntary agreements	

2 Complexity-induced risk problems	*Risk-informed:* (risk agent and causal chain)	• Characterizing the available evidence
		– Expert consensus-seeking tools:
		  – Delphi or consensus conferencing
		  – Meta-analysis
		  – Scenario construction, etc
		– Results fed into routine operation
			Epistemological discourse

	Robustness-focused: (risk absorbing system)	• Improving buffer capacity of risk target through: – Additional safety factors – Redundancy and diversity in designing safety devices – Improving coping capacity – Establishing high reliability organizations	

Risk managers should thus initiate a broader societal discourse to enable participative decision-making. These discursive measures are aimed at finding appropriate conflict resolution mechanisms capable of reducing the ambiguity to a manageable number of options that can be further assessed and evaluated. The main effort of risk management is hence the organization of a suitable discourse combined with the assurance that all stakeholders and public groups can question and critique the framing of the issue as well as each element of the entire risk chain.

Knowledge Characterization	Management Strategy	Appropriate Instruments	Stakeholder Participation
3 Uncertainty-induced risk problems	*Precaution-based:* (risk agent)	• Using hazard characteristics such as persistence, ubiquity etc as proxies for risk estimates Tools include: – Containment – ALARA (as low as reasonably achievable) and ALARP (as low as reasonably possible) – BACT (best available control technology), etc	Reflective discourse
	Resilience-focused: (risk absorbing system)	• Improving capability to cope with surprises – Diversity of means to accomplish desired benefits – Avoiding high vulnerability – Allowing for flexible responses – Preparedness for adaptation	
4 Ambiguity-induced risk problems	*Discourse-based:*	• Application of conflict resolution methods for reaching consensus or tolerance for risk evaluation results and management option selection – Integration of stakeholder involvement in reaching closure – Emphasis on communication and social discourse	Participative discourse

Table 2.5 provides a summary of these four risk strategies and lists the instruments and tools that are most appropriate for the respective strategy. Again it should be emphasized that the list of strategies and instruments is not exhaustive and can be amended if the case requires it.

Managing Interdependencies

In an interdependent world, the risks faced by any individual, company, region or country depend not only on their own choices but also on those of others. Nor do these entities face one risk at a time: they need to find strategies to deal with a series of interrelated risks that are often ill-defined or outside their control. In the context of terrorism, the risks faced by any given airline, for example, are affected by lax security at other carriers or airports. There are myriad settings that demonstrate similar interdependencies, including many problems in computer and network security, corporate governance, investment in research and vaccination. Because interdependence does not require proximity, the antecedents to catastrophes can be quite distinct and distant from the actual disaster, as in the case of the 11 September 2001 attacks when security failures at Boston's Logan Airport led to crashes at the World Trade Center and the Pentagon and in rural Pennsylvania. The same was true in the case of the August 2003 power failures in the north-eastern US and Canada, where the initiating event occurred in Ohio, but the worst consequences were felt hundreds of miles away. Similarly a disease in one region can readily spread to other areas with which it has contact, as was the case with the rapid spread of SARS from China to its trading partners.

The more interdependencies there are within a particular setting (be this a set of organizational units or companies, a geographical area, or a number of countries) and the more that this setting's entities – or participants – decide not to invest in risk reduction while being able to contaminate other entities, the less incentive each potentially affected participant will have to invest in protection. At the same time, however, each participant would have been better off had all the other participants invested in risk-reducing measures. In other words, weak links may lead to suboptimal behaviour by everyone.[12]

For situations in which participants are reluctant to adopt protective measures to reduce the chances of catastrophic losses due to the possibility of contamination from weak links in the system, a solution might be found in a public–private partnership. This is particularly true if the risks to be dealt with are associated with competing interpretations (ambiguities) as to what type of cooperation is required between different epistemic communities as well as risk management agencies in order to deal with various knowledge and competing value claims. Public–private partnerships also provide an interesting alternative in cases in which perceptions differ strongly and external effects are to be expected.

One way to structure such partnerships is to have government standards and regulations coupled with third-party inspections and insurance to enforce these measures. Such a management-based regulatory strategy will not only encourage the addressees of the regulation, often the corporate sector, to reduce their risks from, for example, accidents and disasters, but equally shifts the locus of decision-making from the government regulatory authority to private companies which are as a result required to do their own planning as to how they will meet a set of standards or regulations (Coglianese and Lazer,

2003). This, in turn, can enable companies to choose those means and measures which are most fit for purpose within their specific environment and, eventually, may lead to a superior allocation of resources compared to more top–down forms of regulation. The combination of third-party inspections in conjunction with private insurance is consequently a powerful combination of public oversight and market mechanisms that can convince many companies of the advantages of implementing the necessary measures to make their plants safer and encourage the remaining ones to comply with the regulation to avoid being caught and prosecuted.

Highly interdependent risks that can lead to stochastic contamination of third parties pose a specific challenge for global risk management (i.e. the management of trans-boundary, international and ubiquitous risks). Due to the often particularly decentralized nature of decision-making in this area, a well-balanced mix of consensual (for example international agreements and standards or gentlemen's agreements), coercive (for example government regulation) and incentive-based (for example emission certificates) strategies is necessary to deal with such risk problems. Again these strategies can be best developed in close – international and trans-national – cooperation between the public and the private sectors.

Stakeholder Involvement and Participation

Our emphasis on governance rather than governments or administrations is meant to underline the importance of the inclusion of stakeholders and public groups within the risk handling process and, consequently, on the establishment of adequate public–private partnerships and participatory processes. In the context of this framework we define *stakeholders* as socially organized groups that are or will be affected by the outcome of the event or the activity from which the risk originates and/or by the risk management options taken to counter the risk. Involving stakeholders is not enough, however. Other groups, including the media, cultural elites and opinion leaders, the non-organized *affected public* and the non-organized *observing public*, all have a role to play in risk governance.

Each decision-making process has two major aspects: what and whom to include on the one hand and what and how to select (closure) on the other (Hajer and Wagenaar, 2003; Stirling, 2004). *Inclusion* and *selection* are therefore the two essential parts of any decision- or policy-making activity. Classic decision analysis still offers formal methods for generating options and evaluating these options against a set of predefined criteria. With the advent of new participatory methods, however, the two issues of inclusion and selection have become more complex and sophisticated than purported in these conventional methods.

The present framework advocates the notion of inclusive governance, in particular with respect to global and systemic risks. First and foremost this means that the four major actors in risk decision-making – *political, business, scientific* and *civil society* players – should jointly engage in the process of framing the problem, generating options, evaluating options and coming to a joint conclusion. This has also been the main recommendation of the EU White Paper on European governance (European Commission, 2001). This document endorses transparency and accountability through formal consultation with multiple actors as a means for the EU to address the various frames of governance issues and to identify culture-sensitive responses to common challenges and problems. Similarly to the actors determining the governance of a political union, it is obvious that the actors participating in risk-related decision-making are guided by particular interests which derive not only from the fact that some of them are risk producers – whereas others are exposed to it – but, equally, from their individual institutional rationale and perspective. Such vested interests require specific consideration and measures so that they are made transparent and, if possible, can be reconciled. Inclusive governance, as it relates to the inclusion part of decision-making, requires that:

- there has been a major attempt to involve representatives of all four actor groups (if appropriate);
- there has been a major attempt to empower all actors to participate actively and constructively in the discourse;
- there has been a major attempt to co-design the framing of the risk problem or the issue in a dialogue with these different groups;
- there has been a major attempt to generate a common understanding of the magnitude of the risk (based on the expertise of all participants) as well as the potential risk management options and to include a plurality of options that represent the different interests and values of all parties involved;
- there has been a major effort to conduct a forum for decision-making that provides equal and fair opportunities for all parties to voice their opinion and to express their preferences; and
- there has been a clear connection between the participatory bodies in decision-making and the political implementation level. (Trustnet, 1999; Webler, 1999; Wynne, 2002)

If these conditions are met, evidence shows that actors, along with developing faith in their own competence, use the opportunity and start to place trust in each other and have confidence in the process of risk management (Kasperson et al, 1999; Viklund, 2002; Beierle and Cayford, 2002). This is particularly true for the local level, where the participants are familiar with each other and have more immediate access to the issue (Petts, 1997). Reaching consensus and building trust on highly complex and transgressional subjects such as global change is, however, much more difficult. Being inclusive and open to social groups does not guarantee, therefore, constructive cooperation by those who are invited to participate. Some actors may reject the framing of the issue and choose to withdraw. Others may benefit from the collapse of an inclusive governance process. It is essential to monitor these processes and make sure that particular interests do not dominate the deliberations and that rules can be established and jointly approved to prevent destructive strategizing.

Inclusive governance also needs to address the second part of the decision-making process – reaching closure on a set of options that are selected for further consideration, while others are rejected. Closure here does not mean the final word on a development, risk reduction plan or regulation; rather it represents the product of a deliberation, the agreement that the participants reached. The problem is that the more actors, viewpoints, interests and values are included and thus represented in an arena, the more difficult it is to reach either a consensus or some other kind of joint agreement. A second set of criteria is thus needed to evaluate the process by which closure of debates (be they final or temporary) is brought forth as well as the quality of the decision or recommendation that is generated through the closure procedure.

The first aspect, the quality of the closure process itself, can be subdivided into the following dimensions:

- Have all arguments been properly treated? Have all truth claims been fairly and accurately tested against commonly agreed standards of validation?
- Has all the relevant evidence, in accordance with the actual state-of-the-art knowledge, been collected and processed?
- Was systematic, experiential and practical knowledge and expertise adequately included and processed?
- Were all interests and values considered and was there a major effort to come up with fair and balanced solutions?
- Were all normative judgements made explicit and thoroughly explained? Were normative statements derived from accepted ethical principles or legally prescribed norms?
- Were all efforts undertaken to preserve plurality of lifestyles and individual freedom and to restrict the realm of collectively binding decisions to those areas in which binding rules and norms are essential and necessary to produce the wanted outcome? (Webler, 1995; Demos, 2004)

Turning to the issues of outcome, additional criteria need to be addressed. These have been discussed in the political science and governance literature for a long time (Dryzek, 1994; Rhodes, 1997) and are usually stated as comprising effectiveness, efficiency, accountability, legitimacy, fairness, transparency, acceptance by the public and ethical acceptability. They largely coincide with those that have been postulated earlier for the assessments of risk management options:

- more inclusive procedures enrich the generation of options and perspectives and are therefore more responsive to the complexity, uncertainty and ambiguity of the risk phenomena which are being assessed;
- more rational closure processes provide fairer and socially and culturally more adaptive and balanced judgements;
- the combination of voluntary and regulatory actions in the form of public–private partnerships can be improved through early and constructive involvement procedures; and
- the outcomes derived from these procedures are of higher quality in terms of effectiveness, efficiency, legitimacy, fairness, transparency, public acceptance and ethical acceptability than the outcomes of conventional decision-making procedures.

The potential benefits resulting from stakeholder and public involvement depend, however, on the quality of the participation process. It is not sufficient to gather all interested parties around a table and merely hope for the catharsis effect to emerge spontaneously. In particular, it is essential to treat the time and effort of the participating actors as resources that need to be handled with care and respect (Chess et al, 1998). The participation process should be designed so that the various actors are encouraged to contribute to the process in those areas in which they feel they are competent and can offer something to improve the quality of the final product.

In this respect the four risk classes discussed earlier – simple, complex, high uncertainty and high ambiguity risk problems – support generic suggestions for participation (Renn, 2004b):

1   *Simple risk problems.* For making judgements about simple risk problems a sophisticated approach to involve all potentially affected parties is not necessary. Most actors would not even seek to participate since the expected results are more or less obvious. In terms of cooperative strategies, an *instrumental discourse* among agency staff, directly affected groups (such as product or activity providers and immediately exposed individuals) as well as enforcement personnel is advisable. One should be aware, however, that often risks that appear simple turn out to be more complex, uncertain or ambiguous than originally assessed. It is therefore essential to revisit these risks regularly and monitor the outcomes carefully.

2   *Complex risk problems.* The proper handling of complexity in risk appraisal and risk management requires transparency over subjective judgements and the inclusion of knowledge elements that have shaped the parameters on both sides of the cost–benefit equation. Resolving complexity necessitates a discursive procedure during the appraisal phase with a direct link to the tolerability and acceptability judgement and risk management. Input for handling complexity could be provided by an

epistemological discourse aimed at finding the best estimates for characterizing the risks under consideration. This discourse should be inspired by different science camps and the participation of experts and knowledge carriers. Whether they come from academia, government, industry or civil society, their legitimacy to participate is their claim to bring new or additional knowledge to the negotiating table. The goal is to resolve cognitive conflicts. Exercises such as Delphi, Group Delphi and consensus workshops would be most suitable to serve the goals of an epistemological discourse (Webler et al, 1991; Gregory et al, 2001).

3   *Risk problems due to high unresolved uncertainty.* Characterizing risks, evaluating risks and designing options for risk reduction pose special challenges in situations of high uncertainty about the risk estimates. How can one judge the severity of a situation when the potential damage and its probability are unknown or highly uncertain? In this dilemma, risk managers are well advised to include the main stakeholders in the evaluation process and ask them to find a consensus on the extra margin of safety in which they would be willing to invest in exchange for avoiding potentially catastrophic consequences. This type of deliberation, known as *reflective discourse*, relies on a collective reflection about balancing the possibilities for over- and under-protection. If too much protection is sought, innovations may be prevented or stalled; if we go for too little protection, society may experience unpleasant surprises. The classic question 'How safe is safe enough?' is replaced by the question 'How much uncertainty and ignorance are the main actors willing to accept in exchange for some given benefit?'. It is recommended that policy-makers, representatives of major stakeholder groups and scientists take part in this type of discourse. The reflective discourse can take different forms: round tables, open space forums, negotiated rule-making exercises, mediation, or mixed advisory committees including scientists and stakeholders (Amy, 1983; Perrit, 1986; Rowe and Frewer, 2000).

4   *Risk problems due to high ambiguity.* If major ambiguities are associated with a risk problem, it is not enough to demonstrate that risk regulators are open to public concerns and address the issues that many people wish them to take care of. In these cases the process of risk evaluation needs to be open to public input and new forms of deliberation. This starts with revisiting the question of proper framing. Is the issue really a risk problem or is it in fact an issue of lifestyle and future vision? The aim is to find consensus on the dimensions of ambiguity that need to be addressed in comparing risks and benefits and balancing the pros and cons. High ambiguities require the most inclusive strategy for participation since not only directly affected groups but also those indirectly affected have something to contribute to the debate. Resolving ambiguities in risk debates requires a *participative discourse*, a platform where competing arguments, beliefs

and values are openly discussed. The opportunity for resolving these conflicting expectations lies in the process of identifying common values, defining options that allow people to live their own vision of a 'good life' without compromising the vision of others, to find equitable and just distribution rules when it comes to common resources, and to activate institutional means for reaching common welfare so all can reap the collective benefits instead of a few (coping with the classic commoners' dilemma).[13] Available sets of deliberative processes include citizen panels, citizen juries, consensus conferences, ombudspersons, citizen advisory commissions and similar participatory instruments (Dienel, 1989; Fiorino, 1990; Durant and Joss, 1995; Armour, 1995; Applegate, 1998).

Categorizing risks according to the quality and nature of available information thereon may, of course, be contested among the stakeholders. Who decides whether a risk issue can be categorized as simple, complex, uncertain or ambiguous? It is possible that no consensus may be reached as to where to locate a specific risk. In such cases, a detailed worst-case analysis of possibilities of monitoring and surveillance may constitute the only achievable compromise (reversible removal of risk sources etc, timely detection of adverse effects, strength of surveillance systems). The best means, however, to deal with this conflict is to provide for stakeholder involvement when allocating the different risks into the four categories. This task can be located in the phase of screening as the third component of pre-assessment. Allocating risks to the four categories needs to be done before the assessment procedures start. Over the course of further analysis of risks and concerns the categorization may change – new data and information is collected that may necessitate a reordering of the risk. Yet the risk governance system that is proposed here builds upon the need to classify risks at the beginning and allocate them to different routes of appraisal, characterization, evaluation and management. It seems prudent to have a screening board perform this challenging task. This board should consist of members of the risk and concern assessment team, risk managers and key stakeholders (such as industry, NGOs and representatives of related regulatory or governmental agencies). The type of discourse required for this task is called *design discourse* and is aimed at selecting the appropriate risk and concern assessment policy, defining priorities in handling risks, organizing the appropriate involvement procedures, and specifying the conditions under which the further steps of the risk-handling process will be conducted.

Figure 2.3 provides an overview of the different requirements for participation and stakeholder involvement for the four classes of risk problems and the design discourse. As is the case with all classifications, this scheme shows an extremely simplified picture of the involvement process; indeed it has been criticized for being too rigid in its linking of risk characteristics (complexity, uncertainty and ambiguity) and specific forms of discourse and dialogue (van

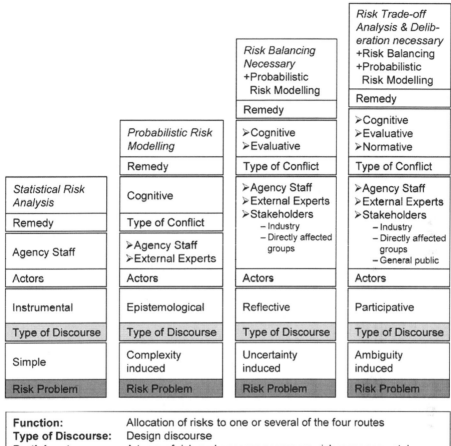

Figure 2.3 The risk management escalator and stakeholder involvement
(from simple via complex and uncertain to ambiguous phenomena)

Asselt, 2005). In addition to the generic distinctions shown in Figure 2.3, it may, for instance, be wise to distinguish between participatory processes based on risk agent or risk-absorbing issues. To conclude these caveats, the purpose of this scheme is to provide general orientation and explain a generic distinction between ideal cases rather than to offer a strict recipe for participation.

Risk Communication

Given the arguments about risk perception and stakeholder involvement, effective communication has to be at the core of any successful activity to assess and manage risks. The field of risk communication initially developed as a means of investigating how expert assessments could best be communicated to the public so that the tension between public perceptions and expert judgement could be bridged. In the course of time this original objective of educating the public about risks has been modified and even reversed as the professional risk community realized that most members of the public refused to become 'educated' by the experts but rather insisted that alternative positions and risk management practices should be selected by the professional community in their attempts to reduce and manage the risks of modern technology (Plough and Krimsky, 1987).

In a recent review of risk communication, William Leiss identified 'three phases in the evolution of risk communication practices' (Leiss, 1996, p85ff). The first phase of risk communication emphasizes the necessity to convey probabilistic thinking to the general public and to educate the layperson to acknowledge and accept the risk management practices of the respective institutions. The most prominent instrument of risk communication in phase I was the application of risk comparisons. If anyone was willing to accept x fatalities as a result of voluntary activities, he or she should be obliged to accept another voluntary activity with fewer than x fatalities. However, this logic failed to convince audiences: people were unwilling to abstract from the context of risk-taking and the corresponding social conditions and they also rejected the reliance on expected values as the only benchmarks for evaluating risks. When this attempt at communication failed, phase II was initiated. This emphasized persuasion and focused on public relations efforts to convince people that some of their behaviour was unacceptable (such as smoking and drinking) since it exposed them to high risk levels, whereas public worries and concerns about many technological and environmental risks (such as nuclear installations, liquid gas tanks or food additives) were regarded as overcautious due to the absence of any significant risk level. This communication process resulted in some behavioural changes at the personal level: many people started to quit some unhealthy habits. However, it did not convince a majority of these people that the current risk management practices for most of the technological facilities and environmental risks were indeed the politically appropriate response to risk. The one-way communication process of conveying a message to the public in carefully crafted, persuasive language produced little effect. Most respondents were appalled by this approach or simply did not believe the message, regardless of how well it was packaged, so phase III evolved. This current phase of risk communication stresses a two-way communication process in which not only are members of the public expected to engage in a social learning process, but so are the risk managers. The objective of this communication effort is to build mutual trust by responding to the concerns of the public and relevant stakeholders. The ultimate goal of risk communication is to assist stakeholders in understanding the rationale of risk assessment results and risk management decisions, and to help them arrive at a balanced judgement that reflects the factual evidence about the matter at hand in relation to their own interests and values (OECD, 2002). Good practices in risk communication help stakeholders to make informed choices about matters of concern to them and to create mutual trust (Hance et al, 1988; Lundgren, 1994).

Risk communication is needed throughout the whole risk handling chain, from the framing of the issue to the monitoring of risk management impacts. The precise form of communication needs to reflect the nature of the risks under consideration, their context and whether they arouse, or could arouse, societal concern. Communication has to be a means to ensure that:

- those who are central to risk framing, risk appraisal or risk management understand what is happening, how they are to be involved, and, where appropriate, what their responsibilities are; and
- others outside the immediate risk appraisal or risk management process are informed and engaged.

The first task of risk communication – facilitating an exchange of information among risk professionals – has often been underestimated in the literature. A close communication link between risk/concern assessors and risk managers, particularly in the phases of pre-assessment and tolerability/acceptability judgement, is crucial for improving overall governance. Similarly, cooperation among natural and social scientists, close teamwork between legal and technical staff, and continuous communication between policy-makers and scientists are all important prerequisites for enhancing risk management performance. This is particularly important for the initial screening phase, where the allocation of risks is performed.

The second task – that of communicating risk appropriately to the outside world – is also a very challenging endeavour. Many representatives of stakeholder groups and, particularly, members of the affected and non-affected public are often unfamiliar with the approaches used to assess and manage risks and/or pursue a specific agenda, trying to achieve extensive consideration of their own viewpoints. They face difficulties when asked to differentiate between the potentially dangerous properties of a substance (hazards) and the risk estimates that depend on the properties of the substance, the exposure to humans and the scenario of its uses (Morgan et al, 2002). Also complicating communication is the fact that some risks are acute, with severe effects that are easy to recognize, whereas others exert adverse effects only weakly but over a long period of time. Yet other risks' effects only start to show after an initial delay. Finally, it is no easy task to convey possible synergies of exposures to industrial substances with other factors that relate to lifestyle (for example nutrition, smoking or the use of alcohol).

Effective communication, or the non-existence thereof, has a major bearing on how well people are prepared to face and cope with risk. Limited knowledge of, and involvement in, the risk management process can lead to inappropriate behaviour in emergency or risk-bearing situations (for example, when facing a pending flood or handling contaminated food or water). There is also the risk of failed communication: consumers or product users may misread or misunderstand risk warnings or labels so that they may, through ignorance, expose themselves to a larger risk than necessary. This is particularly prevalent in countries with high rates of illiteracy and unfamiliarity with risk-related terms. Providing understandable information to help people cope with risks and disasters is, however, only one function of risk communication. Most risk communication analysts list four major functions:

1   *Education and enlightenment*: inform the audience about risks and the handling of these risks, including risk and concern assessment and management.
2   *Risk training and inducement of behavioural changes*: help people cope with risks and potential disasters.
3   *Creation of confidence in institutions responsible for the assessment and management of risk*: give people the assurance that the existing risk governance structures are capable of handling risks in an effective, efficient, fair and acceptable manner (such credibility is crucial in situations in which there is a lack of personal experience and people depend on neutral and disinterested information). It should be kept in mind, however, that trust cannot be produced or generated, but only accumulated by performance, and that it can be undermined by the lack of respect for an individual within such an institution.
4   *Involvement in risk-related decisions and conflict resolution*: give stakeholders and representatives of the public the opportunity to participate in the risk appraisal and management efforts and/or be included in the resolution of conflicts about risks and appropriate risk management options. (Morgan et al, 1992; OECD, 2002)

For all four functions, risk communication needs to address the following topics:

• explaining the concept of probability and stochastic effects;
• explaining the difference between risk and hazard;
• dealing with stigmatized risk agents or highly dreadful consequences (such as nuclear waste or cancer);
• coping with long-term effects;
• providing an understanding of synergistic effects with other lifestyle factors;

- addressing the problem of remaining uncertainties and ambiguities;
- coping with the diversity of stakeholders and parties in the risk appraisal and management phase; and
- coping with inter-cultural differences within pluralist societies and between different nations and cultures.

Although risk communication implies a stronger role for risk professionals to provide information to the public rather than vice versa, it should be regarded as a mutual learning process in line with the requirements that Leiss postulated for phase III. Concerns, perceptions and experiential knowledge of the targeted audience or audiences should thus guide risk professionals in their selection of topics and subjects: it is not the task of the communicators to decide what people need to know but to respond to the questions of what people want to know (the 'right to know' concept, see Baram, 1984). Risk communication requires professional performance both by risk and communication experts. Scientists, communication specialists and regulators are encouraged to take a much more prominent role in risk communication, because effective risk communication can make a strong contribution to the success of a comprehensive and responsible risk management programme.

Notes

1 I am indebted to Gene Rosa for giving me guidance on keeping a healthy balance between a relativist and realist version of risk. For further reading refer to Rosa (1998). It should be noted that this conceptual contribution takes no stand on the controversial issue of *constructivism versus realism of evidence and values* (this topic is extensively reviewed in Mayo and Hollander, 1991, specific positions in Bradbury, 1989; Douglas, 1990; Shrader-Frechette, 1991 and 1995; Wynne, 1992; Laudan, 1996; Jasanoff, 2004). Whether the evidence collected represents human ideas about reality or depicts representations of reality is of no importance for the distinction between evidence and values that is suggested here. Handling risks will inevitably be directed by evidence claims (What are the causes and what are the effects?) and normative claims (What is good, acceptable and tolerable?) It is true that providing evidence is always contingent on existing normative axioms and social conventions. Likewise, normative positions are always enlightened by assumptions about reality (Ravetz, 1999). The fact that evidence is never value-free and that values are never void of assumptions about evidence does not compromise the need for a functional distinction between the two. For handling risks one is forced to distinguish between what is likely to be expected when selecting option x rather than option y on the one hand, and what is more desirable or tolerable: the consequences of option x or option y on the other. It is therefore highly advisable to maintain the classic distinction between evidence and values and also to affirm that justifying claims for evidence versus values involves different routes of legitimization and validation.

2 The Codex Alimentarius Commission was created in 1963 by the Food and Agriculture Organization (FAO) of the UN and the World Health Organization to develop food standards, guidelines and related texts such as codes of practice under the Joint FAO/WHO Food Standards Programme. The main purposes of this programme are protecting health of the consumers and ensuring fair trade

practices in the food trade, and promoting coordination of all food standards work undertaken by international governmental and non-governmental organizations. See www.codexalimentarius.net/web/index_en.jsp.

3   It should also be noted that early warning may of course also benefit from 'non-systematic' findings and incidental/accidental reporting.

4   There are many tools available to model epistemic uncertainty. The Dutch guidance document on uncertainty assessment and communication lists the following tools: sensitivity analysis, error propagation methods, Monte Carlo Analysis, NUSAP (numeral, unit, spread, assessment, pedigree), expert elicitation, scenario analysis, PRIMA (pluralistic framework of integrated uncertainty management and risk analysis) and checklists for model quality assistance (van den Sluijs et al, 2004).

5   With respect to risk and decision-making the term ambiguity has been used with various meanings. Some analysts refer to ambiguity as the conflicting goals of participants in the process (Skinner, 1999), others use the term ambiguity when they refer to the inability to estimate probabilities of an event occurring (Gosh and Ray, 1997; Ho et al, 2002; Stirling, 2003). In the context of the present framework ambiguity denotes the variability in interpretation and normative implications with respect to accepted evidence.

6   The 'drawers' cannot be treated in detail here since this would exceed the scope of this volume (more information in Streffer et al, 2003, p269ff).

7   A first version of the document was released in late 2004. Reports about practical experiences regarding its implementation are not yet available.

8   A similar decomposition has been proposed by the UK government (Environment Agency, 1998; Pollard et al, 2000).

9   The traffic light model in this context is an illustrative means of mapping risks according to their tolerability or acceptability. The same metaphor has also been used to map the degree of controversy or normative ambiguity, for example in the area of siting mobile base stations (Kemp, 1998; Kemp and Greulich, 2004). The criticism that has been raised against using the traffic light model for addressing opposition to base stations is not relevant to the application of this model in the context of risk characterization and evaluation.

10   The terms robustness and resilience have different meanings in different contexts. In most of the *natural hazard literature*, robustness is one of the main components of resilience. In much of the *cybernetic literature*, robustness refers to the insensitivity of numerical results to small changes, while resilience characterizes the insensitivity of the entire system against surprises. Our suggestion for distinguishing the two comes close to the cybernetic use of the terms.

11   The link between precaution and irreversibility was also mentioned in the aforementioned latest report on risk management by the UK Treasury Department (UK Treasury Department, 2004).

12   For a more formal game theoretic treatment of this problem see Kunreuther and Heal (2003).

13   For a more detailed analysis of participatory methods for reaching consensus refer to Barber (1984), Webler (1999) or Jaeger et al (2001).

References

Amy, D. J. (1983) 'Environmental mediation: An alternative approach to policy stalemates', *Policy Sciences*, no 15, pp345–365

Applegate, J. (1998) 'Beyond the usual suspects: The use of citizens advisory boards in environmental decision making', *Indiana Law Journal*, no 73, p903

Armour, A. (1995) 'The citizen's jury model of public participation', in O. Renn, T.

Webler and P. Wiedemann (eds) *Fairness and Competence in Citizen Participation: Evaluating New Models for Environmental Discourse*, Kluwer, Dordrecht and Boston, pp175–188

Baram, M. (1984) 'The right to know and the duty to disclose hazard information', *American Journal of Public Health*, vol 74, no 4, pp385–390

Barber, B. (1984) *Strong Democracy. Participatory Politics for a New Age*, University of California Press, Berkeley, US

Beck, U. (1994) 'The reinvention of politics: Towards a theory of reflexive modernization', in U. Beck, A. Giddens and S. Lash (eds) *Reflexive Modernization. Politics, Tradition and Aesthetics in the Modern Social Order*, Stanford University Press, Stanford, US, pp1–55

Beierle, T. C. and Cayford, J. (2002) *Democracy in Practice: Public Participation in Environmental Decisions*, Resources for the Future, Washington

Boholm, A. (1998) 'Comparative studies of risk perception: A review of twenty years of research', *Journal of Risk Research*, vol 1, no 2, pp135–163

Bradbury, J. A. (1989) 'The policy implications of differing concepts of risk', *Science, Technology, and Human Values*, vol 14, no 4 (fall), pp380–399

Brehmer B. (1987) 'The psychology of risk', in W. T. Singleton and J. Howden (eds) *Risk and Decisions*, Wiley, New York, pp25–39

Brown, H. and Goble, R. (1990) 'The role of scientists in risk assessment', *Risk: Issues in Health and Safety*, no VI, pp283–311

Bruijn, J. A. and ten Heuvelhof, E. F. (1999) 'Scientific expertise in complex decision-making processes', *Science and Public Policy*, vol 26, no 3, pp151–161

Burns, W. J., Slovic, P., Kasperson, R. E., Kasperson, J. X., Renn, O. and Emani, S. (1993) 'Incorporating structural models into research on the social amplification of risk: Implications for theory construction and decision making', *Risk Analysis*, vol 13, no 6, pp611–623

Chess, C., Dietz, T. and Shannon, M. (1998) 'Who should deliberate when?', *Human Ecology Review*, vol 5, no 1, pp60–68

Clark, W. (2001) 'Research systems for a transition toward sustainability', *GAIA*, vol 10, no 4, pp264–266

Codex Alimentarius Commission (2001) *Procedural Manual: Twelfth Edition*, UN Food and Agriculture Organization, Rome, www.fao.org/documents/show_cdr.asp?url_file=/DOCREP/005/Y2200E/y2200e07.htm, accessed 11 August 2005

Coglianese, C. (1999) 'The limits of consensus', *Environment*, vol 41, no 28, pp28–33

Coglianese, C. and Lazer, D. (2003) 'Management-based regulation: Prescribing private management to achieve public goals', *Law and Society*, no 37, pp691–730

Cooke, R. M. (1991) *Experts in Uncertainty: Opinion and Subjective Probability in Science*, Oxford University Press, Oxford and New York

Covello, V. T. (1983) 'The perception of technological risks: A literature review', *Technological Forecasting and Social Change*, no 23, pp285–297

Cross, F. B. (1998) 'Facts and values in risk assessment', *Reliability Engineering and Systems Safety*, no 59, pp27–45

Demos (2004) *See-Through Science: Why Public Engagement Needs to Move Upstream*, monograph by J. Wisdon and R. Willis, Demos, London

Dienel, P. C. (1989) 'Contributing to social decision methodology: Citizen reports on technological projects', in C. Vlek and G. Cvetkovich (eds) *Social Decision Methodology for Technological Projects*, Kluwer, Dordrecht and Boston, pp133–151

Douglas, M. (1990) 'Risk as a forensic resource', *DEADALUS*, vol 119, no 4 (fall), pp1–16

Drottz-Sjöberg, B. M. (1991) *Perception of Risk: Studies of Risk Attitudes, Perceptions, and Definitions*, Center for Risk Research, Stockholm

Dryzek, J. S. (1994) *Discursive Democracy. Politics, Policy, and Political Science (Second Edition)*, Cambridge University Press, Cambridge

Durant, J. and Joss, S. (1995) *Public Participation in Science*, Science Museum, London

Environment Agency (1998) *Strategic Risk Assessment: Further Developments and Trials*, R&D Report E70, Environment Agency, London

European Commission (2000) *Scientific Opinions: First Report on the Harmonization of Risk Assessment Procedures*, European Commission/Health and Consumer Protection Directorate General, Brussels

European Commission (2001) *European Governance: A White Paper*, European Commission, Brussels

European Commission (2003) *Final Report on Setting the Scientific Frame for the Inclusion of New Quality of Life Concerns in the Risk Assessment Process*, European Commission, Brussels

Fiorino, D. J. (1990) 'Citizen participation and environmental risk: A survey of institutional mechanisms', *Science, Technology, & Human Values*, vol 15, no 2, pp226–243

Fischhoff, B. (1985) 'Managing risk perceptions', *Issues in Science and Technology*, vol 2, no 1, pp83–96

Fischhoff, B. (1995) 'Risk perception and communication unplugged: Twenty years of process', *Risk Analysis*, vol 5, no 2, pp137–145

Fischhoff, B., Slovic, P., Lichtenstein, S., Read, S. and Combs, B. (1978) 'How safe is safe enough? A psychometric study of attitudes toward technological risks and benefits', *Policy Sciences*, no 9, pp127–152

Functowicz, S. O. and Ravetz, J. R. (1992) 'Three types of risk assessment and the emergence of post-normal science', in S. Krimsky and D. Golding (eds) *Social Theories of Risk*, Praeger, Westport and London, pp251–273

Goodwin, P. and Wright, G. (2004) *Decision Analysis for Management Judgement*, Wiley, London

Gosh, D. and Ray, M. R. (1997) 'Risk, ambiguity and decision choice: Some additional evidence', *Decision Sciences*, vol 28, no 1 (winter), pp81–104

Graham, J. D. and Rhomberg, L. (1996) 'How risks are identified and assessed', in H. Kunreuther and P. Slovic (eds) *Challenges in Risk Assessment and Risk Management*, The Annals of the American Academy of Political and Social Science, Sage, Thousand Oaks, US, pp15–24

Graham, J. D. and Wiener, J. B. (1995) *Risk vs. Risk*, Harvard University Press, Cambridge, US

Greeno, J. L. and Wilson, J. S. (1995) 'New frontiers in environmental, health and safety management', in R. Kolluru, S. Bartell, R. Pitblade and S. Stricoff (eds) *Risk Assessment and Management Handbook. For Environmental, Health and Safety Professionals*, McGraw-Hill, New York, NY

Gregory, R. S. (2004) 'Valuing risk management choices', in T. McDaniels and M. J. Small (eds) *Risk Analysis and Society. An Interdisciplinary Characterization of the Field*, Cambridge University Press, Cambridge, pp213–250

Gregory, R., McDaniels, T. and Fields, D. (2001) 'Decision aiding, not dispute resolution: A new perspective for environmental negotiation', *Journal of Policy Analysis and Management*, vol 20, no 3, pp415–432

Grossi, P. and Kunreuther, H. (eds) (2005) *Catastrophe Modeling: A New Approach to Managing Risk*, Springer, New York

Hajer, M. und Wagenaar, H. (2003) *Deliberative Policy Analysis: Understanding Governance in the Network Society*, Cambridge University Press, Boston, US

Hammond, J., Keeney, R. and Raiffa, H. (1999) *Smart Choices: A Practical Guide to Making Better Decisions*, Harvard Business School Press, Cambridge, US

Hampel, J. and Renn, O. (eds) (2000) *Gentechnik in der Öffentlichkeit: Wahrnehmung und Bewertung einer umstrittenen Technologie (Second Edition)*, Campus, Frankfurt/Main, Germany

Hance, B. J., Chess, C. and Sandman, P. M. (1988) *Improving Dialogue with Communities: A Risk Communication Manual for Government*, Environmental Communication Research Programme, Rutgers University, New Brunswick, Canada

Hattis, D. (2004) 'The conception of variability in risk analyses: Developments since 1980', in T. McDaniels and M. J. Small (eds) *Risk Analysis and Society: An Interdisciplinary Characterization of the Field*, Cambridge University Press, Cambridge, pp15–45

Hattis, D. and Kennedy, D. (1990) 'Assessing risks from health hazards: An imperfect science', in T. S. Glickman and M. Gough (eds) *Readings in Risk*, Resources for the Future, Washington, pp156–163

Ho, J. L. L., Keller, R. and Keltyka, P. (2002) 'Effects of probabilistic and outcome ambiguity on managerial choices', *Journal of Risk and Uncertainty*, vol 24, no 1, pp47–74

HSE (2001) *Reducing Risk – Protecting People*, Health and Safety Exccutive, London

Hsee, C. and Kunreuther, H. (2000) 'The affection effect in insurance decisions', *Journal of Risk and Uncertainty*, no 20, pp141–159

IAEA (1995) *Guidelines for Integrated Risk Assessment and Management in Large Industrial Areas*, Technical Document IAEA-TECDOC PGVI-CIJV, International Atomic Energy Agency, Vienna

IEC (1993) *Guidelines for Risk Analysis of Technological Systems*, Report IEC-CD (Sec) 381 issued by the Technical Committee QMS/23, European Community, Brussels

IPCS and WHO (2004) *Risk Assessment Terminology*, World Health Organization, Geneva

Jaeger, C., Renn, O., Rosa, E. and Webler, T. (2001) *Risk, Uncertainty and Rational Action*, Earthscan, London

Jasanoff, S. (2004) 'Ordering knowledge, ordering society', in S. Jasanoff (ed) *States of Knowledge: The Co-Production of Science and Social Order*, Routledge, London, pp31–54

Kahneman, D. and Tversky, A. (1979) 'Prospect theory: An analysis of decision under risk', *Econometrica*, vol 47, no 2, pp263–291

Kasperson, J. X., Kasperson, R. E., Pidgeon, N. F. and Slovic, P. (2003) 'The social amplification of risk: Assessing fifteen years of research and theory', in N. F. Pidgeon, R. K. Kasperson and P. Slovic (eds) *The Social Amplification of Risk*, Cambridge University Press, Cambridge, pp13–46

Kasperson, R. E., Golding, D. and Kasperson, J. X. (1999) 'Risk, trust and democratic theory', in G. Cvetkovich and R. Löfstedt (eds) *Social Trust and the Management of Risk*, Earthscan, London, pp22–41

Kasperson, R. E., Jhaveri, N. and Kasperson, J. X. (2001) 'Stigma and the social amplification of risk: Toward a framework of analysis', in J. Flynn, P. Slovic and H. Kunreuther (eds) *Risk Media and Stigma*, Earthscan, London, pp9–27

Kasperson, R. E., Renn, O., Slovic, P., Brown, H. S., Emel, J., Goble, R., Kasperson, J. X. and Ratick, S. (1988) 'The social amplification of risk. A conceptual framework', *Risk Analysis*, vol 8, no 2, pp177–187

Keeney, R. (1992) *Value-Focused Thinking: A Path to Creative Decision Making*, Harvard University Press, Cambridge, US

Keeney, R. and McDaniels, T. (2001) 'A framework to guide thinking and analysis regarding climate change policies', *Risk Analysis*, vol 21, no 6 (December), pp989–1000

Kemp, R. (1998) 'Modern strategies of risk communication: Reflections on recent experience', in R. Matthes, J. Bernhardt and M. Repacholi (eds) *Risk Perception, Risk Communication and its Application to EMF Exposure*, ICNRP 5/98, International Commission on Non-Ionising Radiation Protection and World Health Organization, Geneva, pp117–125

Kemp, R. and Greulich, T. (2004) *Communication, Consultation, Community: MCF Site Deployment Consultation Handbook*, Mobile Carriers Forum, Melbourne, Australia

Klinke, A. and Renn, O. (2002) 'A new approach to risk evaluation and management: Risk-based, precaution-based and discourse-based management', *Risk Analysis*, vol 22, no 6 (December), pp1071–1094

Kolluru, R. V. (1995) 'Risk assessment and management: A unified approach', in R. Kolluru, S. Bartell, R. Pitblade and S. Stricoff (eds) *Risk Assessment and Management Handbook For Environmental, Health, and Safety Professionals*, McGraw-Hill, New York, NY

Kunreuther, H. and Heal, G. (2003) 'Interdependent security', *Journal of Risk and Uncertainty*, Special Issue on Terrorist Risks, vol 26, nos 2/3 (March/May), pp231–249

Kunreuther, H., Novemsky, N. and Kahneman, D. (2001) 'Making low probabilities useful', *Journal of Risk and Uncertainty*, no 23, pp103–120

Laudan, L. (1996) 'The pseudo-science of science? The demise of the demarcation problem,' in L. Laudan (ed) *Beyond Positivism and Relativism: Theory, Method and Evidence*, Westview Press, Boulder, US, pp166–192

Lave, L. (1987) 'Health and safety risk analyses: Information for better decisions', *Science*, no 236, pp291–295

Leiss, W. (1996) 'Three phases in risk communication practice', in 'Annals of the American Academy of Political and Social Science', Special Issue, H. Kunreuther and P. Slovic (eds) *Challenges in Risk Assessment and Risk Management*, Sage, Thousand Oaks, US, pp85–94

Liberatore, A. and Funtowicz, S. (2003) 'Democratizing expertise, expertising democracy: What does this mean, and why bother?', *Science and Public Policy*, vol 30, no 3, pp146–150

Loewenstein, G., Weber, E., Hsee, C. and Welch, E. (2001) 'Risk as feelings', *Psychological Bulletin*, no 127, pp267–86

Löfstedt, R. E. (1997) *Risk Evaluation in the United Kingdom: Legal Requirements, Conceptual Foundations, and Practical Experiences with Special Emphasis on Energy Systems*, working paper no 92, Akademie für Technikfolgenabschätzung, Stuttgart, Germany

Löfstedt, R. E. and Vogel, D. (2001) 'The changing character of regulation. A comparison of Europe and the United States', *Risk Analysis*, vol 21, no 3, pp393–402

Lundgren, R. E. (1994) *Risk Communication: A Handbook for Communicating Environmental, Safety, and Health Risks*, Battelle Press, Columbus, Ohio, US

Mayo, D. G. and Hollander, R. D. (eds) (1991) *Acceptable Evidence: Science and Values in Risk Management*, Oxford University Press, Oxford and New York

Morgan, M. G. (1990) 'Choosing and managing technology-induced risk', in T. S. Glickman and M. Gough (eds) *Readings in Risk*, Resources for the Future, Washington, pp17–28

Morgan, M. G., Fischhoff, B., Bostrom, A. and Atman, C. J. (2002) *Risk Communication: A Mental Models Approach*, Cambridge University Press, Boston and New York

Morgan, M. G. and Fischhoff, B., Bostrom, A., Lave, L. and Atman, C. (1992) 'Communicating risk to the public', *Environmental Science and Technology*, vol 26, no 11, pp2049–2056

Morgan, M. G., Henrion, M. (1990) *Uncertainty: A Guide to Dealing with Uncertainty in Quantitative Risk and Policy Analysis*, Cambridge University Press, Cambridge

National Research Council, Committee on Risk and Decision Making (1982) *Risk and Decision Making: Perspectives and Research*, National Academy Press, Washington

National Research Council, Committee on the Institutional Means for Assessment of Risks to Public Health (1983) *Risk Assessment in the Federal Government: Managing the Process*, National Academy of Sciences, National Academy Press, Washington

OECD (2002) *Guidance Document on Risk Communication for Chemical Risk Management*, OECD, Paris

OECD (2003) *Emerging Systemic Risks: Final Report to the OECD Futures Project*, OECD, Paris

Olin, S., Farland, W., Park, C., Rhomberg, L., Scheuplein, R., Starr, T. and Wilson, J. (1995) *Low Dose Extrapolation of Cancer Risks: Issues and Perspectives*, ILSI Press, Washington

Perritt, H. H. (1986) 'Negotiated rulemaking in practice', *Journal of Policy Analysis and Management*, vol 5 (spring), pp482–95

Petts, J. (1997) 'The public–expert interface in local waste management decisions: Expertise, credibility, and process', *Public Understanding of Science*, vol 6, no 4, pp359–381

Pidgeon, N. F. (1998) 'Risk assessment, risk values and the social science programme: Why we do need risk perception research', *Reliability Engineering and System Safety*, no 59, pp5–15

Pidgeon, N. F. and Gregory, R. (2004) 'Judgment, decision making and public policy,' in D. Koehler and N. Harvey (eds) *Blackwell Handbook of Judgment and Decision Making*, Blackwell, Oxford, pp604–623

Pidgeon, N. F, Hood, C. C., Jones, D. K. C., Turner, B. A. and Gibson, R. (1992) 'Risk perception', in Royal Society Study Group *Risk Analysis, Perception and Management*, The Royal Society, London, pp89–134

Pinkau, K. and Renn, O. (1998) *Environmental Standards: Scientific Foundations and Rational Procedures of Regulation with Emphasis on Radiological Risk Management*, Kluwer, Dordrecht and Boston

Plough, A. and Krimsky, S. (1987) 'The emergence of risk communication studies: Social and political context', *Science, Technology, and Human Values*, no 12, pp78–85

Pollard, S. J. T., Duarte-Davidson, R., Yearsley, R., Twigger-Ross, C., Fisher, J., Willows, R. and Irwin, J. (2000) *A Strategic Approach to the Consideration of 'Environmental Harm'*, The Environment Agency, Bristol, UK

Ravetz, J. (1999) 'What is post-normal science', *Futures*, vol 31, no 7, pp647–653

Renn, O. (1997) 'Three decades of risk research: Accomplishments and new challenges', *Journal of Risk Research*, vol 1, no 1, pp49–71

Renn, O. (2004a) 'Perception of risks', *The Geneva Papers on Risk and Insurance*, vol 29, no 1, pp102–114

Renn, O. (2004b) 'The challenge of integrating deliberation and expertise: Participation and discourse in risk management', in T. L. MacDaniels and M. J. Small (eds) *Risk Analysis and Society: An Interdisciplinary Characterization of the Field*, Cambridge University Press, Cambridge, pp289–366

Rhodes, R. A. W. (1997) *Understanding Governance: Policy Networks, Governance, Reflexivity and Accountability*, Open University Press, Buckingham, UK

RISKO (2000) 'Mitteilungen für Kommission für Risikobewertung des Kantons Basel-Stadt: Seit 10 Jahren beurteilt die RISKO die Tragbarkeit von Risiken', *Bulletin*, vol 3 (June), pp2–3

Rohrmann, B. and Renn, O. (2000) 'Risk perception research – An introduction', in O. Renn and B. Rohrmann (eds) *Cross-Cultural Risk Perception: A Survey of Empirical Studies*, Kluwer, Dordrecht and Boston, pp11–54

Rosa, E. A. (1998) 'Metatheoretical foundations for post-normal risk', *Journal of Risk Research*, no 1, pp15–44

Ross, L. D. (1977) 'The intuitive psychologist and his shortcomings: Distortions in the attribution process', in L. Berkowitz (ed) *Advances in Experimental Social Psychology*, vol 10, Random House, New York, pp173–220

Rowe, G. and Frewer, L. (2000) 'Public participation methods: An evaluative review of the literature', *Science, Technology and Human Values*, no 25, pp3–29

Shome, N., Cornell, C. A., Bazzurro, P. and Carballo, J. E. (1998) 'Earthquakes, records and nonlinear responses', *Earthquake Spectra*, vol 14, no 3 (August), pp469–500

Shrader-Frechette, K. S. (1991) 'Reductionist approaches to risk', in D. G. Mayo and R. D. Hollander (eds) *Acceptable Evidence: Science and Values in Risk Management*, Oxford University Press, Oxford and New York, pp218–248

Shrader-Frechette, K. S. (1995) 'Evaluating the expertise of experts', *Risk: Health, Safety & Environment*, no 6, pp115–126

Sjöberg, L. (1999) 'Risk perception in Western Europe', *Ambio*, vol 28, no 6, pp543–549

Skinner, D. (1999) *Introduction to Decision Analysis* (second edition), Probabilistic Publishers, London

Slovic, P. (1987) 'Perception of risk', *Science*, no 236, pp280–285

Slovic, P. (1992) 'Perception of risk: Reflections on the psychometric paradigm', in S. Krimsky and D. Golding (eds) *Social Theories of Risk*, Praeger, Westport and London, pp153–178

Slovic, P., Finucane, E., Peters, D. and MacGregor, R. (2002) 'The affect heuristic', in T. Gilovich, D. Griffin and D. Kahneman (eds) *Intuitive Judgment: Heuristics and Biases*, Cambridge University Press, Boston, US, pp397–420

Slovic, P., Fischhoff, B. and Lichtenstein, S. (1982) 'Why study risk perception?', *Risk Analysis*, no 2 (June), pp83–94

Stern, P. C. and Fineberg, V. (1996) *Understanding Risk: Informing Decisions in a Democratic Society*, National Research Council, Committee on Risk Characterization, National Academy Press, Washington

Stirling, A. (1998) 'Risk at a turning point?', *Journal of Risk Research*, vol 1, no 2, pp97–109

Stirling A. (2003) 'Risk, uncertainty and precaution: Some instrumental implications from the social sciences', in F. Berkhout, M. Leach, I. Scoones (eds) *Negotiating Change,* Edward Elgar, London, pp33–76

Stirling, A. (2004) 'Opening up or closing down: Analysis, participation and power in the social appraisal of technology', in M. Leach, I. Scoones and B. Wynne (eds) *Science and Citizens Globalization and the Challenge of Engagement*, Zed, London, pp218–231

Streffer, C., Bücker, J., Cansier, A., Cansier, D., Gethmann, C. F., Guderian, R., Hanekamp, G., Henschler, D., Pöch, G., Rehbinder, E., Renn, O., Slesina, M. and Wuttke, K. (2003) *Environmental Standards. Combined Exposures and Their Effects on Human Beings and Their Environment*, Springer, Berlin

Stricoff, R. S. (1995) 'Safety risk analysis and process safety management: Principles and practices', in R. Kolluru, S. Bartell, R. Pitblade and S. Stricoff (eds) *Risk Assessment and Management Handbook: For Environmental, Health, and Safety Professionals,* McGraw-Hill, New York

Thompson, M., Ellis, W. and Wildavsky, A. (1990) *Cultural Theory*, Westview Press, Boulder, US

Trustnet (1999) *A New Perspective on Risk Governance*, Document of the Trustnet Network, Paris, www.trustnetgovernance.com, accessed 11 August 2005

Tversky, A. and Kahneman, D. (1974) 'Judgement under uncertainty. Heuristics and biases', *Science*, no 85, pp1124–1131

Tversky, A. and Kahneman, D. (1981) 'The framing of decisions and the psychology of choice', *Science*, no 211, pp453–458

UK Treasury Department (2004) *Managing Risks to the Public: Appraisal Guidance*, Draft for Consultation, HM Treasury Press, London, www.hm-treasury.gov.uk, accessed 11 August 2005

US Environmental Protection Agency (1997) *Exposure Factors Handbook*, NTIS PB98-124217, EPA, Washington, http://cfpub.epa.gov/ncea/cfm/recordisplay.cfm?deid

=12464, accessed 11 August 2005

van Asselt, M. B. A. (2000) *Perspectives on Uncertainty and Risk*, Kluwer, Dordrecht and Boston

van Asselt, M. B. A. (2005) 'The complex significance of uncertainty in a risk area', *International Journal of Risk Assessment and Management*, vol 5, no 2–4, pp125–158, available at www.inderscience.com/browse/ index.php?journalID =24&year=2005&vol=5&issue=2/3/4, accessed 30 November 2006

van der Sluijs, J. P., Janssen, P. H. M., Petersen, A. C., Kloprogge, P., Risbey, J. S., Tuinstra, W. and Ravetz, J. R. (2004) *RIVM/MNP Guidance for Uncertainty Assessment and Communication: Tool Catalogue for Uncertainty Assessment*, Report No NWS-E-2004-37, Copernicus Institute for Sustainable Development and Innovation and Netherlands Environmental Assessment Agency, Utrecht and Bilthoven, The Netherlands

van der Sluijs, J. P., Risbey, J. S., Kloprogge, P., Ravetz, J. R., Funtowicz, S. O., Corral Quintana, S., Guimaraes Pereira, A., De Marchi, B., Petersen, A. C., Janssen, P. H. M., Hoppe, R. and Huijs, S. W. F. (2003), *RIVM/MNP Guidance for Uncertainty Assessment and Communication*, Report No NWS-E-2003-163, Copernicus Institute for Sustainable Development and Innovation and Netherlands Environmental Assessment Agency, Utrecht and Bilthoven, The Netherlands

Viklund, M. (2002) *Risk Policy: Trust, Risk Perception, and Attitudes*, Stockholm School of Economics, Stockholm

Viscusi, W. K. (1994) 'Risk–Risk Analysis', *Journal of Risk and Uncertainty*, no 8, pp5–18

Vogel, D. (2003) 'Risk regulation in Europe and in the United States', in H. Somsen (ed) *Yearbook of European Environmental Law*, vol 3, Oxford University Press, Oxford

von Winterfeldt, D. and Edwards, W. (1984) 'Patterns of conflict about risk debates', *Risk Analysis*, no 4, pp55–68

WBGU (Wissenschaftlicher Beirat der Bundesregierung Globale Umweltveränderungen) (2000) *World in Transition: Strategies for Managing Global Environmental Risks*, Springer, Berlin

Webler, T. (1995) 'Right discourse in citizen participation. An evaluative yardstick', in O. Renn, T. Webler and P. Wiedemann (eds) *Fairness and Competence in Citizen Participation: Evaluating New Models for Environmental Discourse*, Kluwer, Dordrecht and Boston, pp35–86

Webler, T. (1999) 'The craft and theory of public participation: A dialectical process', *Risk Research*, vol 2, no 1, pp55–71

Webler, T., Levine, D., Rakel, H. and Renn, O. (1991) 'The group Delphi: A novel attempt at reducing uncertainty', *Technological Forecasting and Social Change*, no 39, pp253–263

Wiener, J. B. (1998) 'Managing the iatrogenic risks of risk management', *Risk: Health Safety & Environment*, vol 9, pp39–83

Wynne, B. (1992) 'Risk and social learning: Reification to engagement', in S. Krimsky and D. Golding (eds) *Social Theories of Risk*, Praeger, Westport, US, pp275–297

Wynne, B. (2002) 'Risk and environment as legitimatory discourses of technology: Reflexivity inside out?', *Current Sociology*, vol 50, no 30, pp459–477

3

Wider Governance Issues

Ortwin Renn

The framework covered in the previous two chapters, concerning the areas of risk framing (pre-assessment), appraisal (including risk assessment as well as the assessment of risk-related concerns and the non-physical secondary implications of risk), characterization/evaluation, management and communication, concludes our analysis of the classic components of handling risks. Looking at organizational capacity opens a new set of wider risk governance issues which relate to the interplay between the governing actors and their capability to fulfil their role in the risk governance process.

In discussing the different components of risk appraisal and management, it was implicitly assumed that society has developed the institutional and organizational capability to perform all the tasks prescribed in each component – preferably in a matter-of-fact, objective manner. This is, of course, an ideal picture that masks the realities of the *political* context in which risk governance takes place. In particular, the framing of risk is exposed to many institutional and political forces who may wish to jump on the bandwagon of public dissent or media hype in order to push their own interests (Shubik, 1991). Given the potential of risk perceptions to mobilize public outrage and, thus, to make it impossible for decision-makers not to listen, some actors in society may have an interest in orchestrating 'risk events', while others may have a major motivation for concealing risks or downplaying their impacts. Most political systems have responded to this manoeuvring by establishing independent risk assessment and sometimes management agencies, expecting that these are less likely to be influenced by public pressures. As the European Commission's White Paper on European governance pointed out, the key ingredients of 'good' governance in this sense are openness, participation, accountability, effectiveness and coherence (European Commission, 2001, p10). These requirements are important for all countries but, in particular, for many transitional and most developing countries.

For the analysis of institutional capacity it is useful to distinguish between *assets*, *skills* and *capabilities* (Paquet, 2001). Assets form the social capital for risk governance in the form of knowledge bases and structural conditions for effective management. Skills refer to the quality of institutional and human performance in exploring, anticipating and dealing with existing and emerging risks. Capabilities describe the institutional framework necessary to translate assets and skills into successful policies. These three components constitute the backbone of institutional capacity for risk governance.

Assets include the following:

- *rules, norms and regulations*: these establish rights and obligations. In the risk area, the existence of norms, standards, best practices, legal instruments and so forth has always been a major and often contentious issue, hence the importance of such assets. This is true not only with regard to their prescribing how to deal with risk but also for the absence, or the lack of observance, of rules (for example with regard to the end use of new technologies), which itself constitutes an increasing factor of risk;
- *resources*: these are not limited to financial resources but also comprise an appropriate physical infrastructure for managing risk as well as the availability of adequate information, including the means for information gathering and processing;
- *competencies and knowledge*: this involves providing the necessary education and training and establishing and maintaining a pool of experience and expertise. Education should not only be directed at specialists, but should reach out to the general public, building a culture of risk awareness and prevention; and
- *organizational integration*: the capacity to access and retrieve, in a combination tailored to individual cases, each of these first three types of assets. Organizational integration is a key element, without which otherwise worthy assets will struggle to achieve much.

Using an analogy from mathematics, the three first assets are additive while organizational integration is a multiplying factor. A non-existent organizational capability for integration would nullify the efficacy of the other factors.

Skills are related to the capacity of organizations and institutions to deal with evolving, sometimes chaotic, external conditions. Such conditions should not be considered as an eventuality that cannot be dealt with, but should instead be viewed as input parameters to the risk process that require adequate treatment. Skills should enable political, economic and civic actors to use effectively, and enhance the impact of, the available assets. They relate to:

- *flexibility*: new ways to make sense of a dynamic situation – adapting to change, which in many cases means fighting against established practices and institutional inertia. An example to illustrate this point can be found in the current concern that city planning frequently still follows 19th century practices while the increase in magnitude and frequency of extreme climatic events associated with climate change should dictate a new approach;

- *vision*: bringing new practices into a context that would not naturally generate them – anticipating change. This implies devoting more attention to advanced methodological approaches, such as foresight and scenario planning, and a preparedness to think 'outside the box'; and
- *directivity*: reframing the whole perception of the way of life – driving change that impacts on the outside world rather than limiting oneself to preventing or mitigating the effects of external forces. Several environmental policies (for example the ban on CFCs) and security policies (for example the ban on weapons of mass destruction) adopted at the international level reflect this approach.

Using the same mathematical analogy, the three factors constituting the skills are in an additive relationship with each other. Within that relationship they can exhibit different intensities as a function of the nature of external forces.

Capabilities, finally, constitute the framework in which assets enriched by skills can be exploited for developing and implementing successful risk governance policies. Capabilities can be conceptualized as a structure with several successive layers (Wolf, 2005):

- *relations* link users and sources of knowledge as well as those carrying the authority and those bearing the risk, notably civil society. As previously stated, the participation of civil society in risk governance is essential. Relations should thus be based on inclusive decision making in order to alleviate, at the outset, any circumstances that generate dispute and conflict and consequently aggravate risk;
- *networks* constitute, in terms of structures, a close cooperative structure that goes beyond relations. Halfway between self-organization and hierarchy, networks determine close links between and among groups of principally equal actors; and
- *regimes* establish the rules of the game, the framework in which the actors should act. Both relations and networks are essential for forming and sustaining regimes.

Drawing on the mathematical analogy again, the factors constituting the capabilities are additive, each having a separate but complementary function in the overall build-up of capabilities.

In a world where *human capital* – and in particular brainpower combined with inspiration, courage and a strong ability towards implementation – has largely become the life-blood of society's progress and prosperity, it is quite evident that one of the major keys to the successful handling of risk lies in people's heads. Given the often systemic and global (trans-boundary, international and ubiquitous) nature of today's major risks, special *competencies* and *knowledge* are required. Specialized in-depth expert knowledge in a restricted area or sector may no longer suffice to understand and counteract risks which spread across the boundaries of academic disciplines and business sectors, have several layers of effects, and are determined by a multitude of often interlinked factors. However, compartmentalized specialization is what many educational systems still foster. This approach should, in fact, be replaced by one which emphasizes risk appraisal and management in education at all levels and which considers risk under a broad and multidisciplinary perspective. There is a particular need for this in the engineering, architecture and design disciplines, where a primarily technical focus should be extended to health, safety and environmental risk. Such a new approach, fostering a 'bird's eye perspective' with regard to risk, should be anchored in national science and education policies and should grow to become part of our scientific and technological culture.

All three factors – assets, skills and capabilities – are important variables when assessing and investigating risk governance structures in different countries or risk domains; they can also serve as guiding principles for identifying and researching deficiencies and providing assistance to improve capacity. It may even be possible, based on the above mathematical analogies, to construct an overall performance indicator that could help countries to evaluate their risk governance capacities and to use these elements as pathfinders for establishing new institutional frameworks to achieve improved structures for coping with risk.

The Role of Political Culture

When considering the wider environment of risk handling in modern societies, many classes of influential factors come into play. Only a few can be mentioned here. For example, the distinction between horizontal and vertical governance as introduced in the first chapter of this book can be helpful in describing and analysing cases of risk handling in different countries and contexts (Zürn, 2000). In addition, the interplay between economic, political, scientific and civil society actors needs to be addressed when looking beyond just governmental or corporate actions.

Each country and, in many instances, different risk domains within a country pursue different pathways for dealing with risk. The multitude of risk classification documents and meta-analyses of risk taxonomies is obvious proof of the plurality of risk handling processes and conceptual approaches. It may thus be helpful to search for some underlying principles of these approaches and classify them accordingly.

This exercise of finding common denominators in cultural and national diversity is less of a challenge than one may assume at first glance. Most analysts agree that many of the cognitive factors that govern risk perception are similar throughout the world (Rohrmann and Renn, 2000). In addition, risk management styles are also becoming increasingly homogeneous as the world becomes more globalized (Löfstedt and Vogel, 2001). In spite of the distinct cultural differences between nations and the variations with respect to educational systems, research organizations and structures of scientific institutions, assessment and management of risks and concerns have become universal enterprises in which nationality, cultural background or institutional setting play a minor role only. This is particularly due to the role of science in proposing and justifying regulatory standards. Research establishments and universities have evolved into multinational and cosmopolitan institutions that speak identical or at least similar languages and exchange ideas on worldwide communication networks.

Risk management depends, however, not only on scientific input. It rather rests on three components: *systematic knowledge*, *legally prescribed procedures* and *social values*. Even if the same knowledge is processed by different risk management authorities, the prescriptions for managing risk may differ in many aspects (for example with regard to inclusion and selection rules, interpretative frames and action plans for dealing with evidence). National culture, political traditions and social norms furthermore influence the mechanisms and institutions for integrating knowledge and expertise in the policy arenas. Policy analysts have developed a classification of governmental styles that address these aspects and mechanisms. While these styles have been labelled inconsistently in the literature, they refer to common procedures in different settings (O'Riordan and Wynne, 1987). These are summarized in Table 3.1.

The *adversarial approach* is characterized by an open forum in which different actors compete for social and political influence in the respective policy arena. The actors in such an arena use and need scientific evidence to support their position. Policy-makers pay specific attention to formal proofs of evidence because their decisions can be challenged by social groups on the basis of insufficient use or negligence of scientific knowledge. Risk management and communication is essential for risk regulation in an adversarial setting because stakeholders demand to be informed and consulted. Within this socio-political context, stakeholder involvement is mandatory.

In the *fiduciary approach*, the decision-making process is confined to a group of patrons who are obliged to make the 'common good' the guiding principle of their action. Public scrutiny and involvement of the affected public are alien to this approach. The public can provide input to and arguments for the patrons but are not allowed to be part of the negotiation or policy formulation process. The system relies on producing faith in the competence and the fairness of the patrons involved in the decision-making process. Advisers are selected according to national prestige or personal affiliations. In this political context, stakeholder involvement may even be regarded as a sign of weakness or a diffusion of personal accountability.

Table 3.1 *Characteristics of policy-making styles*

Style	Characteristics	Risk Management
1 *Adversarial approach*	• open to professional and public scrutiny • need for scientific justification of policy selection • precise procedural rules • oriented towards producing informed decisions by plural actors	• main emphasis on mutual agreements on scientific evidence and pragmatic knowledge • integration of adversarial positions through formal rules (due process) • little emphasis on personal judgement and reflection on the side of the risk managers • stakeholder involvement essential for reaching communication objectives
2 *Fiduciary approach (patronage)*	• closed circle of 'patrons' • no public control, but public input • hardly any procedural rules • oriented towards producing faith in the system	• main emphasis on enlightenment and background knowledge through experts • strong reliance on institutional in-house 'expertise' • emphasis on demonstrating trustworthiness • communication focused on institutional performance and 'good record'
3 *Consensual approach*	• open to members of the 'club' • negotiations behind closed doors • flexible procedural rules • oriented towards producing solidarity with the club	• reputation most important attribute • strong reliance on key social actors (also non-scientific experts) • emphasis on demonstrating social consensus • communication focused on support by key actors
4 *Corporatist approach*	• open to interest groups and experts • limited public control but high visibility • strict procedural rules outside of negotiating table • oriented towards sustaining trust to the decision-making body	• main emphasis on expert judgement and demonstrating political prudence • strong reliance on impartiality of risk information and evaluation • integration by bargaining within scientifically determined limits • communication focused on fair representation of major societal interests

The *consensual approach* is based on a closed circle of influential actors, a 'club', who negotiate behind closed doors. Social groups and scientists work together to reach a predefined goal. Controversy is not present and conflicts are reconciled on a one-to-one basis before formal negotiations take place. Risk communication in this context serves two major goals: it is supposed to reassure the public that the club acts in the best interest of the public good and to convey the feeling that the relevant voices have been heard and adequately considered. Stakeholder participation is only required to the extent that the club needs further insights from the affected groups or that the composition of the club is challenged.

The *corporatist approach* is similar to the consensual approach but is far more formalized. Well-known experts are invited to join a group of carefully selected policy-makers representing the major forces in society (such as employers, unions, churches, professional associations and environmentalists). Similar to the consensual approach, risk communication is mainly addressed to outsiders: they should gain the impression that the club is open to all 'reasonable' public demands and that it is trying to find a fair compromise between public protection and innovation. Often the groups represented within the club are asked to organize their own risk management and communication programmes as a means of enhancing the credibility of the whole management process.

Although these four styles cannot be found in pure form in any country, they form the backdrop of socio-political context variables against which specific risk governance structures are formed and operated. These structures, along with the individual actors' goals and the institutional perspectives they represent, would need more specific attention and, for the time being, are difficult to classify further.

Conclusions

A prototype version of this framework is outlined in the present volume (and summarized in Figure 5.1, page 97). The framework has been designed on the one hand to include enough flexibility to allow its users to do justice to the wide diversity of risk governance structures, and on the other to provide sufficient clarity, consistency and unambiguous orientation across a range of different risk issues and countries.

Part I of this volume has discussed a comprehensive risk handling chain, breaking down its various components into three main phases: pre-assessment, appraisal and management. The two intermediate and closely linked stages of risk characterization and evaluation have been placed between the appraisal and management phases and can be assigned to either of them, depending on the circumstances: if the interpretation of evidence is the guiding principle for characterizing risks, then risk and concern assessors are probably the most appropriate people to handle this task; if the interpretation of underlying values and the selection of yardsticks for judging acceptability are the key problems, then risk managers should be responsible. In an ideal setting, however, this task of determining a risk's acceptability should be performed in a joint effort by both assessors and managers. At any rate, a comprehensive, informed and value-sensitive risk management process requires a systematic compilation of results from risk assessment, risk perception studies and other context-related aspects as recommended and subsumed under the category of risk appraisal. Risk managers are thus well advised to include all the information related to the risk appraisal in evaluating the tolerability of risks and in designing and evaluating risk reduction options. The crucial task of risk communication runs parallel to all phases of handling risk: it assures transparency, public oversight and mutual understanding of the risks and their governance.

Wider governance issues have also been addressed. The starting point here was the observation that collective decisions about risks result from an inter-action between science communities, governmental or administrative actors, corporate actors and actors from civil society at large. The interplay of these actors has been discussed with reference to public participation, stakeholder involvement and governance structures (horizontal and vertical). In addition, the need for appropriate organizational capacity as a prerequisite for effective risk governance has been highlighted and a typology of regulatory styles provided. These variables also co-determine the institutional structure, the processing of information and values, and the quality of the outcome in terms of regulations or management options.

References

European Commission. (2001) *European Governance: A White Paper*, EU, Brussels

Löfstedt, R. E. and Vogel, D. (2001) 'The changing character of regulation: A compar-ison of Europe and the United States', *Risk Analysis*, vol 21, no 3, pp393–402

O'Riordan, T. and Wynne, B. (1987) 'Regulating environmental risks: A comparative perspective', in P. R. Kleindorfer and H. C. Kunreuther (eds) *Insuring and Managing Hazardous Risks: From Seveso to Bhopal and Beyond*, Springer, Berlin, pp389–410

Paquet, G. (2001) 'The New Governance, subsidiarity, and the strategic state,' in OECD (ed) *Governance in the 21st Century*, OECD, Paris, pp183–215

Rohrmann, B. and Renn, O. (2000) 'Risk perception research – An introduction', in O. Renn and B. Rohrmann (eds) *Cross-Cultural Risk Perception. A Survey of Empirical Studies*, Kluwer, Dordrecht and Boston, pp11–54

Shubik, M. (1991) 'Risk, society, politicians, scientists, and people', in M. Shubik (ed) *Risk, Organizations, and Society*, Kluwer, Dordrecht and Boston, pp7–30

van der Sluijs, J. P., Janssen, P. H. M., Petersen, A. C., Kloprogge, P., Risbey, J. S., Tuinstra, W. and Ravetz, J. R. (2004) *RIVM/MNP Guidance for Uncertainty Assessment and Communication: Tool Catalogue for Uncertainty Assessment*, Report No NWS-E-2004-37, Copernicus Institute for Sustainable Development and Innovation and Netherlands Environmental Assessment Agency, Utrecht and Bilthoven, The Netherlands

Webler, T. (1999) 'The craft and theory of public participation: A dialectical process', *Risk Research*, vol 2, no 1, pp55–71

Wolf, K. D. (2005) 'Civil society and the legitimacy of governance beyond the state and empirical explorations', in A. Benz and I. Papadopoulos (eds) *Governance and Democratic Legitimacy: Transnational, European, and Multi-Level Issues*, Routledge, London

Zürn, M. (2000) 'Democratic governance beyond the nation-state: The EU and other international institutions', *European Journal of International Relations*, vol 6, no 2, pp183–221

Part II

Experiences

The 'Experiences' section of this volume is practice-oriented. It is primarily designed for practitioners of regulatory agencies and industry, who often feel that they need to make 'acceptability' and 'tolerability' decisions without clear methodological guidance. This section focuses mainly on empirical attempts to define and apply tolerability of risk models. In this volume, the expression 'Tolerability of risk models' refers to experiences involving not only conceptual creativity, but also followed by practical implementation.

This section therefore looks into the implementation side of the tolerability of risk issue, exploring frameworks designed to help reach the tough decisions about those risks that, as a society, we decide to broadly accept, to tolerate and reduce, or to refuse. The 'Tolerability of Risk' (ToR) model developed by the HSE in the UK offers the most comprehensive approach to date. For this reason, the ToR framework will be discussed extensively in Chapters 4, 5 and 6 of Part II . Chapter 4 of this volume presents the historical development that resulted in the formulation of the HSE model. It helps assess the extent to which the original ideas that gave birth to the framework can be attributed to specific circumstances and how much they may result from more universal sources of inspiration. Chapter 5 presents the 'regulator's story' behind ToR, helping practitioners to envisage both practical opportunities and limitations. The prerequisites of a well-functioning ToR framework are presented in Chapter 6 and could be a source of inspiration for translating the model into new areas of regulation.

The purpose of Part II is not to 'promote' any specific model but to review opportunities as well as limitations. For this reason, in addition to the specific difficulties encountered by regulators (Chapter 5), Chapter 7 provides a more systematic account of potential limitations of the applicability of the ToR framework to other risk fields.

4

A Historical Perspective
on Tolerability of Risk

Jim McQuaid

Introduction

The reform of the approach to state regulation of health and safety had its foundation in the 1972 report of the Robens Committee of Inquiry (Robens, 1972). The committee was established by government in 1970 to examine the reasons for a perceived failure of state regulation to fulfil the expectations of society with regard to the health and safety performance of industry. The perception was based on a number of features of the industrial sector in the late 1960s:

- health and safety performance appeared to have reached a plateau in terms of recorded incidences of accidents and ill health;
- rapid technological developments and the increasing scale and complexity of industry were imposing unsustainable burdens on the legislative machinery of the state;
- the responsibilities of the many different actors in the system were not clearly distinguished;
- the vulnerability of the public to risks from industrial activities was increasingly demonstrated in incidents both in the UK and abroad; and
- there were particular worries about the hazards of toxic and dangerous substances and the inadequacy of the traditional approach to their control.

How the Law Developed Pre-1970

The situation described above was a consequence of the way the law had developed up to 1970. The Robens Committee examined the history of this development and came to the following main conclusions:

- legislation had emerged piecemeal throughout the 170 years of state intervention;
- each successive statute was aimed at remedying a single ascertained evil;
- the result was a body of detailed law needing constant extension to deal
  with new problems;
- the process of deciding on enactments took excessive time (15 years in one
  quoted instance); and
- there was no overall logic or consistency in the body of law.

The practical, empirical approach worked, but only up to a point; it was
increasingly unable to cope with modern industrial practices and structures.
The Robens Committee was asked to consider whether any changes were
needed in the scope and nature of the enactments, in measures for protecting
the public, and in the extent of voluntary action.

The Robens Committee's Approach and Findings

Over the years, there had been many commissions and committees appointed
to enquire into particular areas or aspects of the subject. There had, however,
never been a comprehensive review of the subject as a whole. The committee
took the far-reaching decision to look at the system and to ask what, if
anything, was wrong with it. The system comprised the whole complex of
arrangements and activities. The committee recognized two broad elements –
regulation and supervision by the state, and individual regulation and self-help.
They concluded that the issue underlying the questions they had been asked to
address lay in the relationship, balance and interaction between these two
elements. They observed that many of their findings could not have emerged
from a partial examination of the field. This is an example of the identification
of the emergent properties of a complex system that are not exhibited by
reductionist analysis of the system's components considered in isolation.

The examination led to the following main findings:

- there was widespread apathy about health and safety at all levels and across
  all sectors of industry;
- there was too much law and it was of the wrong kind, attempting to
  prescribe in detail how industry should manage its activities;
- Parliament was unable to cope with the legislative burden;
- there was built-in obsolescence in many of the enactments; and
- there was fragmentation of administrative jurisdictions with no joined-up
  thinking between them.

Response and Implementation

The Robens Committee's response was to propose a whole new philosophy on which to base a more effective balancing of the respective interests of the state and all those who, in one way or another, influence the creation of industrial risks. The main pillars of this philosophy were that:

- primary responsibility for control should lie with those who create the risks;
- the legislative framework should create conditions for self-regulation – in other words a freedom to devise solutions satisfying safety goals;
- industry should take an active role in setting standards and developing guidance on good practice; and
- health and safety should be a normal part of the management function.

The recommendations of the Robens Committee were implemented in the Health and Safety at Work etc Act (HSWA) of 1974. The 'etc' refers to the inclusion of a duty to protect the public as well as workers. The act places general duties on the main parties and, in effect, provides clarity of responsibilities. These are general in the sense that there is no reference to the level of hazard or risk but the duties are qualified insofar as they are to be achieved 'so far as is reasonably practicable'. The general duties are supported by regulations dealing with specific circumstances and the duties under the regulations are usually (though not always) also qualified by the idea of reasonable practicability. The general duties and those in regulations are amplified by Approved Codes of Practice (ACOPs) and by guidance and advice. These differ in their evidential status. An ACOP must be complied with unless an alternative can be shown to work at least as well, with the onus for doing so being on the duty holder. Guidance and advice has no evidential status and the onus is on the regulator to show why the advice should have been followed in the circumstances. ACOPs and guidance define or represent what is reasonably practicable. Thus the idea of reasonable practicability permeates the legislative framework except where there are incontrovertible risks (for example with certain carcinogenic substances such as asbestos) or where a change in industry practice must be achieved (for example the replacement of railway rolling stock having low crash-worthiness characteristics).

The implementation of the HSWA framework was taken forward by the progressive replacement of prescriptive legislation specifying detailed requirements applying to individual industrial sectors or activities with a minimal structure of regulations of wide scope setting out the ends to be achieved. The regulations were firmly based on risk assessment in the standards they set, relying on scientific evidence and methodological examination of the causation of harm in order to ensure proportionality of response – a key principle of good regulation (Better Regulation Task Force, 2000). They allow for considerable discretion by the duty holder in approaches to dealing with risks and in interpreting what is reasonably practicable where that qualification applies to the statutory duty. Any such interpretation must itself be based on a risk assessment by the duty holder for the particular circumstances in which the risks arise. These include the level of hazard, the people who are affected and the harm to which they are exposed, the applicability of good practice standards established by an authoritative body, and the scope for reasonably practicable improvements to reduce the risks.

The safety philosophy of the legislative framework and its implementation reflects the reality that things are not usually either 'safe' or 'not safe', in the sense that there will always be some level of risk. But they can be made to be 'safe enough' and this equates to compliance with the principle that the risks have been reduced so that they are 'as low as reasonably practicable' (the ALARP principle). Decisions on whether risks are ALARP are in most cases determined by adherence, as a minimum, to authoritative good practice. Where good practice has not been developed or is of doubtful relevance, there is a need to anticipate what might go wrong and to examine the effectiveness, including cost-effectiveness, of options for reducing risks.

The ALARP Principle

The ALARP principle is central to consideration of how far it is necessary to go in pursuing risk reduction. The duty to demonstrate that risks are ALARP amounts to having to argue the case for the safety of the measures in place in terms of the residual risks. Where hazards are low and authoritative good practice applies, the argument will be short and to the point. In high hazard industries, and especially where a 'permissioning regime' applies (HSC, 2003), the argument will be correspondingly more elaborate. Quantification of the risks and the costs of risk reduction can be part of the argument. The overriding consideration is that safety must be given the benefit of any doubt, whether the doubt refers to the applicability of good practice or the treatment of uncertainties in analysing options for reducing the risks. Where the risks are high, the balance in favour of safety must be substantial. Both of these stipulations have been established by case law (Edwards *v.* National Coal Board, 1 All ER 743, 1949). The emphasis in applying the ALARP principle is on weighing up the options for improvement in safety. This process is more judgemental than deterministic, though with the caveat that the reasoning behind the judgements needs to be made explicit and transparent to others. The consequence is that the risk-based approach provides a mechanism for constructive dialogue between the regulator and the regulated, in keeping with the philosophy advocated by the Robens Committee.

The effectiveness of the risk-based legislative framework has been scrutinized on many occasions, most extensively in the *Review of Health and Safety Regulation* (HSC, 1994). The advantages and disadvantages have been examined in *Risk-Based Regulation: Setting Goals for Health and Safety* (Bacon, 1994), and a discussion of experiences and problems in administering the framework is given in *Application of Risk-Based Strategies to Workers' Health and Safety* (Rimmington et al, 2003).

The Tolerability of Risk Framework

By the 1980s, the Health and Safety Executive (HSE), the regulator appointed under the HSWA, had developed considerable competence in the practice of risk assessment. This was mirrored externally by the growth of occupational safety and health as a professional discipline. As a result of this and the increasing evidence base provided by extensive research programmes, the judgemental nature of ALARP decisions in most industrial situations posed few problems that could not be resolved by professionalism and the application of common sense. However, in high hazard industries, the nature of the risks, the potential effects on the public and the high levels of expenditure associated with risk reduction options together led to a need for a unifying philosophy justifying the residual risks involved. It was not sufficient to state the risks and to declare that they were accepted as ALARP on the basis that no further expenditure was justified to reduce them. The question was: What was the meaning and significance of these residual risks? This question came to the fore in the report by Sir Frank Layfield (Layfield, 1987) on the planning inquiry for the Sizewell B nuclear power station. He proposed that the HSE should 'formulate and publish guidelines on the tolerable levels of individual and societal risk to workers and the public from nuclear power stations'.

In response, the HSE developed and published in 1988 the original Tolerability of Risk (ToR) framework for public consultation, and this was revised and published in 1992 (HSE, 1992). The framework has subsequently been generalized for wider application to all industrial risks (HSE, 2001). This general description set the pattern for similar statements on risk decision-making which all government departments were required to publish (Strategy Unit, 2002). All these developments and future extensions are described in other chapters in this volume.

Disclaimer

The discussion in this chapter is based on the author's experiences in HSE and its predecessors. However, the views expressed are the author's own and do not represent official policy.

References

Bacon, J. (1994) *Risk-Based Regulation: Setting Goals for Health and Safety*, conference on Probabilistic Safety Assessment and Management (PSAM-II), San Diego, CA

Better Regulation Task Force (2000) *Principles of Good Regulation*, Cabinet Office, London

HSC (1994) *Review of Health and Safety Regulation*, HSE Books, Sudbury, UK

HSC (2003) *Policy Statement: Our Approach to Permissioning Regimes*, HSE Books, Sudbury, UK

HSE (1988, rev 1992) *The Tolerability of Risk from Nuclear Power Stations*, HSE Books, Sudbury, UK

HSE (2001) *Reducing Risks, Protecting People*, HSE Books, Sudbury, UK

Layfield, Sir F. (1987) *Sizewell B Public Inquiry: Report by Sir Frank Layfield*, HMSO, London

Rimington, J., McQuaid, J. and Trbojevic, V. (2003) *Application of Risk-Based Strategies to Workers' Health and Safety*, Reed Business Information, Doetincham, The Netherlands

Robens Committee (1972) *Safety and Health at Work, Cmnd 5034*, HMSO, London

Strategy Unit (2002) *Risk: Improving Government's Capability to Handle Risk and Uncertainty*, Cabinet Office, London

5

Tolerability of Risk:
The Regulator's Story

Tony Bandle

Introduction

In partnership with its stakeholders, the Health and Safety Executive (HSE) has the task of protecting people's health and safety by ensuring that risks from work activities are properly controlled. In meeting this task, the HSE undertakes a risk-based regulation of work activities that, as this chapter will illustrate, is all about sensible, proportionate risk management which is, itself, essentially a matter of applying sound judgement.

It is no part of the HSE's game plan that risks should be reduced to zero – this cannot be achieved in practice; only the dead can be at zero risk! Thus the HSE, when contemplating the introduction or amendment of regulation in general or when judging whether a particular duty-holder has done enough in the particular case, has to decide when 'safe' is safe enough. In making this decision, the HSE has to consider, given the status quo:

- whether there is anything more duty-holders can do to control risk from their undertaking; and
- if there is, should they be required to do it?

The answer to the first question is a matter of fact, a question of what can be physically achieved in practice given the particular circumstances – what occupational health and safety law refers to as 'practicable'. However, the answer to the second question is further conditioned by the limitation on resources and the need to use them where they will do most good. In this light 'more should be done' only if it constitutes a proportionate response to the need to further control risk – what occupational health and safety law refers to as 'reasonably practicable': we do not want duty-holders wielding hammers to crack nuts; we want them to do only what is necessary to reduce risks to as low as reasonably practicable (ALARP).

But what constitutes a proportionate response? In part, this involves a consideration of the benefits achieved by doing more and of the costs involved – duty-holders should not be asked to spend pots of money to achieve only insignificant improvements in risk control or to gold-plate existing measures. But it also involves making value judgements: what level of risk *ought* duty-holders be required to achieve or – to put it another way – what level of risk is tolerable or acceptable bearing in mind the benefits that are generated by undertaking the hazardous activity in the first place, which, necessarily, involves taking some risk.

In response to these questions, the HSE has established a framework for decision-making with processes, a tool-kit, which enable the HSE to adopt a coherent approach to providing answers in all the multifarious work environments which the HSE oversees (which range in size and complexity from workshops run by one person and their dog to nuclear power stations) and, whereby, the judgements the HSE makes in its risk-based decision-making can be made openly, transparently and accountably, taking heed of the needs, interests and values of society.[1]

The HSE's Tool-Kit

Good practice

For the great majority of decisions about whether 'more should be done', the decision can be made straightforwardly on the grounds of whether there is a source of relevant good practice and good engineering/safety principles to which duty-holders can turn and, in the particular case, whether or not duty-holders are already complying with these.

Good practice and good engineering/safety principles are established and maintained by processes of collective risk assessment and risk management; and they carry authority, representing as they do a consensus between stakeholders as to what is reasonably practicable. Such consensus often reflects a judgement on what is needed to control risks ALARP because much good practice is empirical and/or evolutionary in origin rather than based on quantitative assessment of the risks and control measures.

The HSE regards as authoritative sources of relevant good practice the Health and Safety Commission's (HSC) Approved Codes of Practice and its own guidance documents (naturally!). We may also recognize other sources such as guidance produced by other government departments, standards produced by standards-making bodies (CEN, ISO and so forth) and guidance agreed by a body (for example a trade federation) representing an industrial/occupational sector, providing the practice they set out has been formulated in a way which fits the description in the preceding paragraph. Furthermore, existing good practice must be kept up to date in the light of new technology or new information regarding the hazardous activity to which the practice is intended to be applied.

When good practice is applied in the particular case, the HSE and duty-holders both need to be sure that it is fully relevant to the circumstances, and duty-holders should be prepared to institute additional controls where it is not (for example, the good practice may deal only with on-site risks and there may also be, in the particular case, off-site risks to consider). Duty-holders may also be able to read across from good practice formulated with an industry other than theirs in mind.

The HSE's expectation is that where relevant good practice is available, duty-holders will apply it. The HSE would not normally accept a lower standard of risk control than that provided by current good practice, but duty-holders are at liberty to adopt a different approach to controlling risks – they simply need to demonstrate that the alternative will achieve at least as good a level of control (in other words that risks have been reduced ALARP).

ALARP demonstrations

It is sometimes the case that there is no source of good practice to indicate whether or not 'more should be done', that what there is only provides part of the answer, or that duty-holders wish to proceed by other than the established good practice. In these circumstances, to demonstrate that what they have done or propose to do reduces risks ALARP, duty-holders need to return to first principles and consider explicitly the issues set out above: the costs and benefits of doing more and what level of risk ought to be achieved.

This will require duty-holders to assess the risk, both at the status quo and after the proposed control measure is in place, to determine the benefit achieved in risk reduction, which can then be compared with the cost, and to allow the risk to be measured against risk-tolerability criteria to decide whether it ought to be reduced further. In many cases, circumstances will be such that the assessment can be largely qualitative, relying on professional judgement. However, in some circumstances, more formal assessment is required.

Formal risk assessment

More formal risk assessment may be required where any, some or all of the following apply:

- the risk is judged to be close to being intolerable;
- the technology/process concerned is complex;
- the technology/process concerned is novel;
- there is significant uncertainty; and
- public/political concern is high because, for example, large numbers could be killed in one incident if the risk were realized, there was a chance of

widespread, irreversible damage, or vulnerable groups such as children were potential victims.

Even this more formal assessment will often involve a qualitative process in which professional judgement has the largest role in the decision-making. However, where major investments in health and safety are being made, or formal safety cases are being considered, it is likely that decision-making will depend more on rigorous quantified risk assessment (QRA). Even here, it is the highly structured analysis, needed for the QRA, of the possible pathways to safety failures and the effectiveness of the measures provided to prevent such failures – not the risk numbers – that provide the basis for dialogue and judgement.

Decision-making in the nuclear and on- and off-shore major hazard industries is likely to involve this more rigorous, quantified approach. However, there is still much reliance on good practice – safety codes and standards, many negotiated and adopted internationally. These are valuable aids to simplifying what otherwise might be complex, opaque and time-consuming decision-making, but sometimes they have the disadvantage (particularly in the global market context) of over-constraining negotiations on what is 'safe enough' in the particular instance.

The tolerability of risk and the ToR framework

To make a value judgement about whether more should be done to reduce risk and what the level of risk ought to be, it is necessary to compare that risk against certain risk criteria. Being highly expert and experienced in its field enables the HSE to say with some authority what the level of risk is and how it might best be controlled, but this expertise and experience does not warrant the HSE determining what the risk ought to be – the 'is' does not lead to the 'ought'.

The HSE needs to look not to its own values but, rather, to society's to determine whether a risk is tolerable. To designate a risk as 'tolerable' means that it is not to be regarded as negligible or something that might be ignored but, rather, as something to be kept under review and possibly reduced still further, depending on circumstances. There is also the clear implication that there are some risks that just cannot be tolerated.

Individuals tolerate different levels of risk, depending on the benefits they think they will gain from taking them. Equally, society's tolerance of different risks varies for a whole variety of reasons, some relatively straightforward and amenable to scientific evaluation and others complex expressions of deep-seated, psychological attitudes. It is against this intricate (and largely unquantifiable) background that the HSE has incorporated the concept of 'tolerability of risk' into its decision making and it does this by applying the Tolerability of Risk (ToR) framework.

The ToR framework (Figure 5.1) is constructed to reflect how people in general approach risk, in other words that some risks are so high that they would, ordinarily, be viewed as intolerable whatever the benefits that might be gained by taking them. Other risks are seen as too small to be of any further concern; in between are risks at a level where they are of concern but can be tolerated provided that they are reduced ALARP.

As indicated by the labelling of the vertical axis in Figure 5.1, when locating a risk in the framework, the HSE requires consideration of both the individual risk and the societal concerns engendered by the hazardous activity in question. Individual risk concerns the potential for harm to individuals, through injury or ill-health, for example. Societal concerns relate to harm to the wider community, damage to the social fabric caused by a loss of public trust in the regulatory system, and to the duty-holders creating and controlling the risk, should the risk materialize.

Thus locating a risk in the framework provides a guide as to what more, if anything, ought to be done as regards further risk control. It will inform the HSC or HSE's decision-making when considering risk regulation in general, but where it really bites is in informing HSE decision-making when it considers what particular duty-holders have done about particular risks.

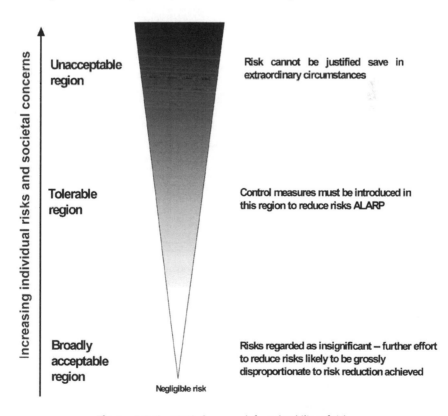

Figure 5.1 The HSE's framework for tolerability of risk

Based on past court rulings,[2] the HSE requires duty-holders to adopt a control measure unless they can show that the cost of so doing is 'grossly disproportionate' to the benefit achieved. Clearly, this means that there may well be some circumstances where duty-holders have to adopt a control measure which involves some degree of disproportion (less than 'gross'). However, the HSE has not formulated an algorithm to calculate the maximum degree of disproportion that it would expect duty-holders to tolerate in any particular circumstances; rather the HSE determines how hard it pushes duty-holders to do more to control a particular risk by – given the status quo – where that risk falls in the ToR framework. For example:

- if the risk falls into the 'unacceptable' region, the HSE will be extremely reluctant to accept *any* arguments from duty-holders for not 'doing more';
- for risks in the 'broadly acceptable' region, the HSE will advise duty-holders to turn their attention to other risks (whilst making sure they maintain the controls which achieved that level of risk in the first place); and
- as regards risks in the 'tolerable' region, how hard the HSE pushes depends on where the risk lies in relation to the other two regions, taking an increasingly hard line as the 'unacceptable' region is approached – in other words the higher the risk, the greater can be the disproportionality between cost and benefit without the HSE conceding it to be 'grossly disproportionate'.

To make the ToR framework into a practical decision-making tool in which particular risks can be located, the regions need to be delineated by tolerability criteria. In the case where the risk is of death to individual workers, currently the HSE designates a risk of 1 in 1000 per annum as the dividing line between what is 'tolerable' and what is 'unacceptable' for any but fairly exceptional circumstances. For individual members of the public who have a risk imposed on them 'in the wider interest', the HSE adopts a limit at a stricter order of magnitude, 1 in 10,000 per annum. At the other end of the spectrum, the HSE believes that an individual risk of death of 1 in 1,000,000 per annum for the public (including workers) corresponds to a very low level of risk and should be considered as 'broadly acceptable'.

These criteria are considered to reflect society's views. The upper boundary was determined by analogy with high-risk industries generally regarded by society as sufficiently well-regulated, and taking account of people's voluntary acceptance of risks in particular situations and the general levels of risk at work that people accept for a personal benefit (such as pay). The lower boundary was arrived at by looking at the level of risks that people seem to regard as not worth worrying about given the background level of risk in which they live.

As regards societal concerns, the HSE has not thus far adopted any criteria apart for those concerns arising out of society's aversion to large-scale accidents: society is, generally, more concerned with one accident killing 10 people than with 10 accidents killing one person, and this aversion gives rise to a public expectation that hazardous activities which could result in major incidents leading to multiple deaths will be risk-managed to a higher standard than the norm. Based on public/parliamentary discussions in the 1980s of the risks posed by the chemical complex at Canvey Island, the HSE employs a criterion whereby the risk of an accident causing the death of 50 people or more in a single event should be regarded as intolerable if the frequency is estimated to be more than 1 in 5000 per annum (see the annex to the following chapter for more details about these calculations).

It must be emphasized that the ToR framework has no formal, legal status – there is no mention of it in the Health and Safety at Work Act or its regulations – and the individual risk and societal risk criteria are in no respect compliance limits. Rather, all provide for the HSE and duty-holders a basis for dialogue, discussion and negotiation on the control strategies and standards of protection that are to be adopted.

Furthermore, the HSE deals with many other risks, in addition to those where death is the likely consequence, where it uses risk tolerability criteria (for example occupational exposure limits) or benchmarks (for example good practice – see above) to judge whether duty-holders should do more or not and which must also reflect society's view. It is clear from the above that the HSE has to have effective, open and transparent processes for engaging its stakeholders in making and explaining judgements about the criteria, which may vary from risk to risk and over time.

Cost–benefit analysis (CBA)

As discussed above, deciding whether 'more should be done' as regards major investments in health and safety or when considering formal safety cases is likely to require the more rigorous approach of QRA. Such a rigorous approach will also be required in respect of the consideration of costs and benefits which forms part of the ALARP demonstration: in other words a CBA will have to be undertaken.

The benefits to be considered are first those associated with the putative reduction in harm to individuals arising from the increased control of risk: in other words the injury and ill-health averted. But benefits may also accrue to the wider community. These may concern putative reduction in tangible harm, such as property damage, or a reduction in the harm to the social fabric caused by a loss of public trust in the regulatory system, and in the duty-holders creating and controlling the risk, should the risk materialize.

The costs are those necessarily incurred as a result of instituting the increased risk control, offset by any cost savings which arise, such as the putative loss of production prevented by the reduction in risk and the actual productivity gains which can be achieved by increased control (for example, automating a process can make it more efficient as well as safer).[3]

As already discussed, part of the decision of whether 'more should be done' involves weighing the costs against the benefits. Conventional cost–benefit analysis usually requires benefits to exceed costs before a measure can be adopted but, as mentioned above, in the field of occupational health and safety the courts have introduced the bias on the side of health and safety of requiring that duty-holders should adopt a 'practicable' control measure unless they can show that the cost of so doing is grossly disproportionate to the benefit achieved.

Enforcement management model

The HSE employs ToR/good practice in judging whether individual duty-holders have done enough to meet their legal obligations: in other words that they have done what is necessary to reduce risks ALARP. A further judgement HSE inspectors will sometimes have to make concerns the enforcement action they should take where duty-holders have not met their legal obligations. If enforcement action is taken, this can involve a range of measures, from the relatively low-key reminding duty-holders of their obligations verbally or by letter, via the issue of formal Improvement Notices and/or Prohibition Notices, to high-profile prosecution in the courts. In practice, none, one or a combination of these measures may be undertaken depending on the circumstances.

Based on ToR/good practice, the HSE has developed a means of helping inspectors decide what would be appropriate enforcement action in particular circumstances, the Enforcement Management Model (EMM).[4] This is not intended to fetter inspectors' discretion when making enforcement decisions or applying direct enforcement in any particular case; rather it is intended to promote consistency and proportionality in enforcement by confirming the parameters that need to be considered and the risk-based criteria against which decisions are made.

Central to the EMM process is a 'risk gap analysis', which measures the actual risk arising from the conditions provided by the duty-holder (in other words where the duty-holder is) and compares it against a benchmark, the level of risk remaining after the duty-holder has complied with legal requirements (in other words where the duty-holder should be), which is determined through ToR/good practice. The EMM uses the size of the gap (extreme, substantial, moderate or nominal) to indicate where appropriate enforcement action might lie in the circumstances.

Future Developments and Challenges

To summarize, good practice, the ToR framework and the tolerability criteria provide the HSE and its stakeholders with an effective framework for decision-making which enables the dialogue, discussion and negotiation on the control strategies and standards of protection to be applied and enforcement action to be taken. However, out of all this a number of further issues arise.

Trade-offs

Discussion around the adoption of control strategies sometimes involves complex judgements about trade-offs, which may involve competing safety claims (for example worker safety vs. public safety or occupational safety vs. environmental safety) or pit safety concerns against such matters as human rights (for example the safety of care workers vs. the human dignity of their patients) or economic/social development (for example as may arise in land-use planning around major hazards sites).

A key challenge for the HSE is how, given its statutorily constrained remit, to achieve trade-offs which not only satisfy its own concerns but also those of other interested parties, particularly as regards other regulators within the context of joined-up government. Where relevant good practice and good engineering principles do not provide benchmarks for when 'safe' is 'safe enough', there may be a temptation to elaborate and refine the ToR framework and/or undertake research to establish a whole range of context-specific criteria. However, far better is to accept the framework/criteria for what they are: a sensible basis for mature dialogue about challenging trade-offs.

The real trick is to create the right culture, conventions and competencies to enable inclusive decision-making which attracts broad support and trust. The work, for example, of Whitehall's cross-government group on implementing the Treasury's 'principles of managing risks to the public'[5] and, in a European context, that of TRUSTNET should make significant contributions to pulling this off.[6]

Health risks

Use of the ToR framework in practice has shown it to be particularly successful in assisting decision-making in relation to acute risks – for example of accidents, particularly those arising in nuclear and on- and off-shore major hazards (after all, it was developed in the first place to handle risks from nuclear power generation) – but less so with chronic risks, for example those to health. The problems arise, primarily, because of the difficulties inherent in the assessment of such chronic risks.

For example, with acute risks, the harmful consequence of the risk being realized is reasonably easy to define – usually death for those hazardous activities where the ToR framework is most likely to be applied. However, deleterious health effects range across a continuum and this raises the question of which end point should be used in the definition of the harm to be assessed. Furthermore, the differences in individual susceptibilities results in much greater uncertainty in the consequences of a chronic risk being realized, and because of latency and threshold effects health risks may not be realized until after the worker exposed has retired. These are some of the difficulties that make the determination of appropriate ToR criteria (and the location of risk in the ToR framework) a much more problematic exercise for chronic risks such as those arising from exposure to toxic chemicals or noise.

However, the discussion so far has assumed the framing of occupational health and safety issues in terms of 'risk'. This is a productive model for the HSC or HSE to use because there exist well-established methods for estimating risk, and a risk-based approach to decision-making can be systematic and analytical, enabling a proportionate and targeted regulatory response based on the degree of risk determined.

But not all occupational health and safety issues fit neatly into the 'risk model'. That model is most useful where regulatory intervention is aimed at getting the duty-holder to prevent or reduce workers' (and others') exposure to the hazard. However, there are issues where prevention/reduction by the duty-holder may not be the only strategy to adopt because the hazard occurs outside the work environment as well as within it. Workers and their health-care advisers may also be able to contribute significantly to remedial processes, and other models besides the 'risk model' – such as those used by health promotion specialists – will also have to be considered. In practice, a combination of models will often prove most fruitful.

Conclusion

It has to be acknowledged that operating the framework for decision-making is not without its demands on the HSE: to date, it has required from the HSE a hands-on and, therefore, resource-intensive approach. There is a significant cost to the HSE in establishing and maintaining the cadre of technically competent inspectors and assessors who involve themselves in the close examination of duty-holders' ALARP demonstrations (which, in the case of the nuclear and on- and off-shore major hazard industries, often involve detailed, quantitative risk assessments) and in the production of good practice which must be formulated to reflect what is reasonably practicable in the light of current information and technology. This burden may ease, however, if the moves towards increased self-regulation, more partnership working and 'earned autonomy' on the part of duty-holders – as promoted in the HSC's current strategy for workplace health and safety (HSC, 2004) – are realized.

One significant positive is that the framework for decision-making is very adaptable and flexible, and the HSE can apply it across the whole range of occupational activities falling under its aegis. What is an opportunity, however, can also be seen by some as a threat. Some duty-holders may prefer the reassurance of more formal, rigid, standards-based approaches (whereby what they need to do is spelled out for them). Reaching a common understanding with duty-holders, and other stakeholders, presents a significant communications challenge to the HSE.

While acknowledging the limitations of the framework, the HSE is confident that it has provided, and will continue to provide, a highly effective, flexible tool for both the HSE and duty-holders to use in deciding what to do about managing risks. In contrast with more formal, standards-based approaches, it enables a ready response to be made to changes in scientific knowledge, risk perceptions and technical advances, but without a requirement on designers to adopt state-of-the-art solutions.

Notes

1   Discussed in detail in HSE's publication *Reducing Risks, Protecting People* (HSE, 2001), commonly referred to as 'R2P2'.
2   The key case is Edwards *v.* The National Coal Board ([1949] 1 KB 704; [1949] 1 All ER (743)).
3   Precisely which costs and benefits are taken into account depends on the context of the decision-making. Two broad categories of risk-based decision-making can be identified:
    i)   the **regulator** (the HSC and HSE and, ultimately, Ministers and Parliament) has to decide on the degree and form of regulatory control for a hazardous activity; or
    ii) the **duty-holder** (mainly employers and the self-employed) has to decide how to meet the regulatory requirements imposed by the regulator according to their own particular circumstances.
    In the case of the duty-holder meeting regulatory requirements, the scope of the decision-making is restricted by the legal framework so that, for example, only the costs incurred by the duty-holder can be counted. Where the HSC and HSE are deciding on the regulatory regime, their not being duty-holders means the scope is not so restricted and, for example, costs to society can also be counted.
4   See www.hse.gov.uk/enforce/emm.pdf.
5   See www.hm-treasury.gov.uk/media//8B2AE/risk_principles_220903.pdf.
6   See www.trustnetgovernance.com/.

References

HSC (2004) *A Strategy for Workplace Health and Safety in Great Britain to 2010 and Beyond*, report, HSC, London, available at www.hse.gov.uk/aboutus/hsc/strategy.htm

HSE (2001) *Reducing Risks, Protecting People*, HSE Books, Sudbury, UK, available at www.hse.gov.uk/risk/theory/r2p2.htm

Applying the HSE's Risk Decision Model: Reducing Risks, Protecting People

Jean-Marie Le Guen

Introduction

In recent years several issues have made the public aware that people and the society in which they live are exposed to numerous risks and that the way in which these risks are managed can affect their lives. Typical issues that have engendered such awareness include:

- *direct threats* – from events such as those that took place in the US on 11 September 2001, malicious and criminal attacks on information technology systems, or potential acts of terrorism involving chemical or biological agents;
- *health and safety concerns* – issues such as BSE, the combined vaccine for measles, mumps and rubella (MMR), travel by rail, flooding, and unhealthy lifestyles leading to problems such as obesity, high blood pressure and diabetes;
- *risks to the environment* – for example emissions causing climate change or pollution of the planet from the use of non-biodegradable products (plastics etc); and
- *ethical considerations* – for example the transfer of risks to countries that may not have the institutional means or safety culture to handle them properly.

This growing realization about risks and their effects has led industry and governments (and regulators acting on their behalf) to reappraise how they assess and manage risks, and citizens to consider how they handle risks in everyday life and to pay closer attention to risks they are prepared to tolerate.

In turn, those who make decisions on risks and their management nowadays are increasingly adopting a deliberative approach for such tasks (OECD, 2001; O'Neill, 2002; Commons Select Committee on Defence, 2002). The overall aim of this approach is to achieve compromises or even consensus with those affected or interested, particularly on how the risks posed by a particular hazard should be addressed, including the choice of the appropriate strategies, criteria and measures for reaching decisions.

This is not easy. Those who take decisions on risks have the duty to ensure that risks are properly controlled and that the measures in place are commensurate to the risks. On the other hand, account has to be taken of public perceptions despite the fact that such perceptions are at least partially driven by biases, questionable evidence, false assumptions and sensation. Moreover, the public is not a homogenous entity. Decision makers are faced with competitive claims, values, opinions from experts, pressure groups and so on.

The HSE's Approach to Decision-Making

The HSE's decision-making framework 'Reducing Risks, Protecting People' tackles the above problems by setting out the ground rules for interacting deliberatively and meaningfully with stakeholders on all aspects of the risks decision-making process. The underlying aim throughout is to ensure that stakeholders will see the whole process as valid and thereby accept the decisions reached.

Many elements of the system adopted by the HSE are similar to those that have been developed by others for informing and reaching decisions (for example the Presidential/Congressional Commission on Risk Assessment and Risk management of 1997). However, the HSE framework goes further by setting out the protocols used for assessing risks, addressing uncertainty and undertaking cost–benefit analysis and by providing the criteria for deciding on what risks are unacceptable, tolerable or broadly acceptable. The system involves the following stages:

1   defining and characterizing the issue;
2   examining the options available for addressing the issue and their merits;
3   adopting a particular course of action for addressing the issue efficiently and in good time, informed by the findings of (1) and (2) above and in the expectation that as far as possible stakeholders will support it;
4   implementing the decisions; and
5   evaluating the effectiveness of actions taken and revisiting the decisions and their implementation if necessary.

These stages are not entirely independent of each other. In practice, progression through the stages is an iterative process and the information or perspectives gathered while progressing from one stage to another often require the revisiting of earlier stages of the process.

The model also requires the involvement of stakeholders at all stages. As already mentioned, the aim of such involvement is to reach a wider consensus on all aspects of the process. The consensus includes an understanding that the decision-maker must be the final arbiter where agreement is not possible, for example because different stakeholders hold opposite views based on deep-rooted beliefs.

The five stages are examined in more detail below.

Stage 1 – Defining and characterizing the issue

Defining the issue

In stage 1, decision-makers consider how the issue can be framed or described in terms of problems to be tackled and the means for tackling them. For example, the storage of hazardous substances at a particular site could be framed as an issue with two quite different dimensions:

- whether adequate safeguards can be put in place for storing the hazardous substance safety; and
- whether the substance should be stored at all at the particular site.

The issue could be framed either way, giving rise to quite different problems. The way an issue is framed has a considerable influence on the measures that should be put in place for addressing the risk. Decision-makers therefore have to pay particular attention to whether their stakeholders agree with the way they have framed the issue.

Characterizing the issue in terms of risk

The framing of the issue may point to it being one where a decision on proportionality of action requires information on the risks. In such cases, decision-makers will need to characterize the risk quantitatively and qualitatively, to describe how it arises and how it impacts on those affected and on society at large. Such information is usually obtained by undertaking an assessment of the risks. Assessing risks involves identifying the hazards associated with the risk issue – in other words what in a particular situation could cause harm or damage – and then assessing the likelihood that harm will actually be experienced by a specified population and what the consequences would be. The scope of the assessment is wide and covers both the risks to people exposed (known as individual risks) and identifying whether the hazard impacts on society as a whole (known as societal risks or societal concerns).

The extent to which each of these issues is considered in the assessment will depend on the nature and attributes of the hazard as well as the context in which people interact with the hazard in question.

For example, many hazards in everyday life are well known, familiar, easy for people to gauge the actual threat they give rise to, have no stigma attached to them and do not cause society any significant concern (for example risks associated with the chlorination of water). In such cases decision-makers are likely to pay more attention to the level of residual individual risks after measures have been introduced rather than to the societal concerns (if any) that they might engender. On the other hand, gauging the extent of the societal concerns caused by a hazard is likely to be a major consideration when considering the control measures that should be introduced for addressing a hazard that is new and unfamiliar, and where its realization would generate a socio-political response (for example the risk associated with genetically modified organisms).

In short, the assessment provides an estimate of the risks by examining available data, experience of harm, scientific information and so on. This estimate, more often than not, will be in the form of a judgement rather than a rigorous numerical value of the risk. The policy process then couples the scientifically-based judgements about risks with policy considerations about the approach to their control. The latter (sometimes separately described as risk evaluation) includes such considerations as the relative weightings to be attached to likelihood and consequence and the way that public perceptions of the risk should be taken into account.

Procedures for addressing uncertainty when assessing risks

In practice, uncertainty permeates the risk assessment process. There are invariably gaps in knowledge that prevent a risk estimate from being determined numerically, For example, there may be a lack of knowledge about the degree of harm that a particular hazard entails, since often the extent to which people are exposed to the hazard, or the susceptibility of one person over another, will vary. Gaps in knowledge are addressed by making judicial assumptions, usually through the adoption of credible scenarios on how the hazards might be realized and what the consequences would be if they were.

Where there is uncertainty about the extent of exposure, risks are usually assessed in relation to a hypothetical person. A 'hypothetical person' describes an individual who is in some fixed relation to the hazard: for example the person most exposed to it, a person living at some fixed point or a person with some assumed pattern of life. For example, occupational exposure to chemicals, entailing adverse consequences after repeated exposure for long periods, is often controlled by considering the exposure of a hypothetical person who is in good health and works exactly 40 hours a week.

To ensure that all significant risks for a particular hazard are adequately covered, there will usually have to be a number of hypothetical persons constructed. For example, for each population exposed to the hazard, there will usually be a hypothetical person specifically constructed for determining the control measures necessary to protect that population.

Relating assessments to a hypothetical person has several advantages. Persons actually exposed to the risks can compare their own circumstances to those associated with the measures deemed necessary to control the risks found for the hypothetical person and decide whether they or their family incur a greater or lesser risk and therefore whether the measures in place are adequate in their circumstances. Decision-makers can also reach similar conclusions in respect of those they have to protect. Moreover, the approach allows all relevant factors to be taken into account in the assessment of the risks (for example human factors, where relevant).

In addition the concept of the hypothetical person has the considerable advantage that it allows the risk of a certain process, activity, situation or so on to be assessed meaningfully and independently of the exposure of persons actually exposed to the risks. This is because in applying the concept, it is assumed that exposure to the hazard is for the time period that was fixed when the credible scenario for the exposure of the hypothetical person was agreed upon.

Uncertainty about the degree of harm attached to a particular hazard is usually tackled by making assumptions about consequences and likelihoods. The credible scenarios can range from a 'most likely' worst case to a 'worst case possible', depending on the degree of uncertainty. For example, by assuming that exposure to a putative carcinogenic chemical will cause cancer the chemical becomes subject to a very stringent control regime. Though such risk assessments made on scenarios are inevitably narrower in scope than a full-blown risk assessment, this may not be a serious limitation if the scenarios are carefully chosen to reflect what could happen in reality. The process for tackling uncertainty in those cases is illustrated in Figure 6.1.

Stage 2 – Examining the options available and their merits

Identifying options

Once the problem has been characterized, decision-makers then identify the options available for managing the risks. These can range from doing nothing to introducing measures to get rid of the cause of the problem altogether or to reduce it to a level which people are prepared to live with so as to secure certain benefits, in the confidence that the risk is worth taking and that it is being properly controlled.

In looking at options, decision-makers are encouraged to examine:

- *good practice* for addressing the hazards identified and evaluating whether it is relevant and sufficient. If specific good practice is not available we would also examine the merits of good practice that applies in comparable circumstances (if decision-makers believe that this is directly transferable or can be suitably modified to address the hazard);

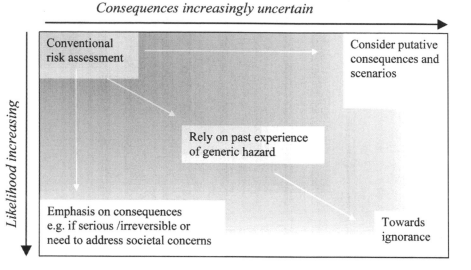

Figure 6.1 Procedures for addressing uncertainty when assessing risks

- *constraints attached to a particular option*: for example whether the option is technically feasible, or whether there are legal constraints on its adoption. The general principle is that the option adopted will improve or at least maintain standards of health, safety and welfare;
- *any adverse consequences associated with a particular option*. Very often adopting an option for reducing one particular risk of concern may create or increase another type of risk. For example, banning a particular solvent may increase the use of a more hazardous one, reducing airborne concentration of substances in the workplace by exhaust ventilation may increase risk in the community or vice versa. Therefore for each option which has adverse consequences the trade-off between reducing the target risk and the increase in other risks must be examined;
- *how much uncertainty is attached to the issue under consideration* and – as a consequence – *the precautionary approach* that should be adopted to ensure that decisions reached are in line with the precautionary principle. The treatment of uncertainty in the HSE framework is intrinsically biased towards ensuring that higher rather than lower standards of health and safety are adopted when reaching decisions. As the degree of uncertainty increases, and depending on certain other characteristics attached to a particular hazard (for example whether the risk, if realized, could result in consequences that are irreversible or could detrimentally affect future generations), there is an increasing shift towards requiring more

stringent measures to mitigate the risks. Moreover, in cases where the benefits cannot justify the risks, the framework requires that consideration is given to banning the activity, process or practice giving rise to the hazard;

- *how far the options succeed in improving (or at least maintaining) standards*;
- the *costs and benefits* attached to each option, by looking at what is required to implement each option and the degree of risk reduction it is likely to achieve. These analyses of costs and benefits (CBA) are performed according to the technical conventions used generally by Government as published by HM Treasury (HM Treasury, 1997); and
- what is the most appropriate *regulatory instrument* in the range available to decision-makers for achieving the objectives for managing the risks in question.

Stage 3 – Adopting decisions

This stage reviews all the information gathered previously with a view to selecting the most appropriate option for managing the risks. Decision-makers usually adopt the criteria developed by the HSE for regulating and managing risks (HSE, 2001a). This is reproduced in Annex 1.[1]

Stage 4 – Implementing the decisions

Stage 4 requires decision-makers to:

- have a plan for taking action by looking ahead and setting priorities for ensuring that risks requiring most attention are tackled first, based on the risk assessment taken in Stage 2;
- set up a system for monitoring and evaluating progress, for example by identifying potential indicators for evaluating how far the control measures introduced have been successful in addressing the problem; and
- comply with well-established principles on the hierarchy of measures for the prevention of risks, for example eliminating risks, combating the risk at source, generally applying sound engineering practice (such as inherently safer design) and applying collective protective measures rather than individual protective measures.

Stage 5 – Evaluating the effectiveness of action taken

Stage 5 completes the process for ensuring that risks are properly managed by establishing procedures to review decisions after a suitable interval to establish:

- whether the actions taken to ensure that the risks are adequately controlled achieved that result;

- whether decisions previously reached need to be modified and, if so, how (for example because levels of protection considered at the time to be good practice may no longer be regarded as such as a result of new knowledge, advances in technology or changes in the level of societal concerns);
- how appropriate was the information gathered in the first two stages of the decision-making process to assist decisions for action, for example the methodologies used for the risk assessment and the cost–benefit analysis (if prepared) or the assumptions made; and
- whether improved knowledge and data would have helped to reach better decisions.

Such evaluations are an ongoing process which decision-makers need to plan carefully. They have to ensure, for example, that there are appropriate systems in place for monitoring and evaluating progress. And they have to get their timing right, since some time might elapse before the full impact of risk reduction measures can be monitored.

Applying the Framework in Practice

The application of the framework can be a time-consuming and resource-intensive process. Though in principle it should be applied to all decisions on risks and their management, in practice decision-makers can adopt several shortcuts. For example, they can:

- take into account that societal concerns are often absent for a wide range of hazards. For example, this is often the case for those hazards that are familiar or where the risks they give rise to are generally accepted as being well controlled. In those circumstances decision-makers need to focus on the individual risks when determining where the hazard falls in the ToR triangle;
- consider that in most circumstances the implementation of authoritative good practice would provide adequate standards of protection; and
- decide to use the procedure set out in the five-stage decision-making process above on those occasions where good practice for the hazards in question has not been identified or is found to result in inadequate control of the risks. This will often be the case for hazards that are new or not well studied, or where the circumstances in which people interface with the hazard are untypical or exceptional.

Though the framework was adopted by the HSE for reaching decisions on risks arising from work activities, it has been used increasingly by other regulators and industry at large for adopting decisions on a wide variety of risks. This has included the management of risks arising from nuclear power stations, the transport of dangerous substances (HSC, 1991), the response to risks of fires (Office of the Deputy PM, 2003), and the rate at which decaying gas pipes are replaced.[2]

Conclusions

All recent studies on the management of risk have found the need for greater openness, transparency and stakeholder involvement in the decision-making process. The deliberative decision-making process adopted by the HSE has a proven record for engendering trust and reaching decisions seen as valid by stakeholders. It achieves this by exposing all stages of the process to public scrutiny in a way that allows stakeholders to understand and judge the appropriateness of the decisions reached in the light of the balance of expert opinion, costs, benefits and preferences of society and with due consideration given to the nature and scale of the risks.

ANNEX

HSE Criteria for Adopting Decisions

An edited version of these is reproduced below.

The criteria are described in the form of a framework (illustrated in Figure 5.1, page 97), known as Tolerability of Risk (ToR). The triangle represents the increasing level of 'risk' for a particular hazardous activity (measured by the individual risk and societal concerns it engenders) as we move from the bottom of the triangle towards the top. The dark zone at the top represents an unacceptable region. For practical purposes, a particular risk falling into that region is regarded as unacceptable whatever the level of benefit associated with the activity. Any activity or practice giving rise to risks falling in that region would be ruled out as a matter of principle, unless the activity or practice can be modified to reduce the degree of risk so that it falls within one of the regions below or there are exceptional reasons for the activity or practice to be retained.

The light zone at the bottom, on the other hand, represents a broadly acceptable region. Risks falling into this region are generally regarded as insignificant and adequately controlled. The HSE, as a regulator, would not usually require further action to reduce risks unless reasonably practicable measures are available. The levels of risk characterizing this region are comparable to those that people regard as insignificant or trivial in their daily lives.

They are typical of the risk from activities that are inherently not very hazardous, or from hazardous activities that can be, and are, readily controlled to produce very low risks. Nonetheless, decision-makers would take into account that often there is a statutory duty imposed on those who create risks to ensure the health and safety of people who may be affected by their activities.

The zone between the unacceptable and broadly acceptable regions is the tolerable region. Risks in that region are typical of the risks from activities that people are prepared to tolerate in order to secure benefits, in the expectation that:

- the nature and level of the risks are properly assessed and the results used properly to determine control measures. The assessment of the risks needs to be based on the best available scientific evidence and, where evidence is lacking, on the best available scientific advice;
- the residual risks are not unduly high and kept ALARP; and
- the risks are periodically reviewed to ensure that they still meet the ALARP criteria, for example by ascertaining whether further or new control measures need to be introduced to take into account changes over time, such as new knowledge about the risk or the availability of new techniques for reducing or eliminating risks.

Benefits for which people generally tolerate risks typically include employment, lower cost of production, personal convenience or the maintenance of general social infrastructure such as the production of electricity or the maintenance of food or water supplies.

Figure 5.1 is a conceptual model. The factors and processes that ultimately decide whether a risk is unacceptable, tolerable or broadly acceptable are dynamic in nature and are sometimes governed by the particular circumstances, time and environment in which the activity giving rise to the risk takes place. For example, standards change, public expectations change with time, what is unacceptable in one society may be tolerable in another and what is tolerable may differ in times of peace or war.

Nevertheless, the protocols, procedures and criteria described in this document should ensure that, in practice, risks are controlled to such a degree that the residual risk is driven down the tolerability range so that it falls either in the broadly acceptable region or is near the bottom of the tolerable region, in keeping with the duty to ensure health, safety and welfare so far as is reasonably practicable.

Tolerability Limits

The ToR framework can in principle be applied to all hazards. When determining the reasonably practicable measures for any particular hazard which gives rise to risks that need to be controlled, the acceptability of the chosen control option depends in part on where the boundaries are set between the unacceptable, tolerable or broadly acceptable regions in Figure 5.1. The choice is generally dictated by the outcome of deliberation and negotiation with stakeholders reflecting the value preferences of stakeholders and the practicability of possible solutions.

When the decision-maker is a regulator, what is unacceptable, tolerable or broadly acceptable in specific circumstances is often spelled out or implied in legislation, approved codes of practice or regulatory guidance documents or reflected in what constitutes good practice; in other words there is no need to set explicit ToR boundaries. For example, on the basis of its wealth of experience accumulated over the years in dealing with its stakeholders, the HSE subscribes as a matter of policy to the following indicative criteria as to where these boundaries lie for risks in a *limited category*, those entailing the *risk of individual or multiple deaths*. These criteria are merely guidelines to be interpreted with common sense and are not intended to be rigid benchmarks requiring compliance in all circumstances. They may, for example, need to be adapted to take account of societal concerns or preferences.

Boundary between the 'broadly acceptable' and 'tolerable' regions for risk-entailing fatalities

The HSE believes that an individual risk of death of 1 in 1,000,000 per annum for both workers and the public corresponds to a very low level of risk and should be used as a guideline for the boundary between the broadly acceptable and tolerable regions. We live in an environment of appreciable risks of various kinds that contribute to a background level of risk – typically a risk of death of 1 in 100 per year, averaged over a lifetime. A residual risk of 1 in 1,000,000 per year is extremely small when compared to this background level of risk. Indeed, many activities which people are prepared to accept in their daily lives for the benefits they bring (for example using gas and electricity or engaging in air travel) entail or exceed such levels of residual risks.

Moreover, many of the activities entailing such a low level of residual risk also bring benefits that contribute to lowering the background level of risks. For example, though electricity kills a number of people every year and entails an individual risk of death in the region of 1 in 1,000,000 per annum, it also saves many more lives, by providing homes with light and heat, operating lifts and life-support machines, and through a myriad of other uses. Indeed, it is the combined effect of many activities involving such low levels of residual risk that contributes to the wealth of the nation and leads to improvements in health and longevity.

Boundary between the 'tolerable' and 'unacceptable' regions for risk-entailing fatalities

The criterion for individual risk for this boundary is not as widely applicable as the one mentioned above for the boundary between the 'broadly acceptable' and 'tolerable' regions. This is because risks may be unacceptable on the grounds of a high level of risk to an exposed individual or because of the repercussions of an activity or event on the wider society. Indeed, it would be quite unusual for high levels of individual risk not to engender societal concerns, on the grounds of fairness, for example.

The converse is not, however, true – society can be seized by hazards that pose, on average, quite low levels of risk to any individual but could impact unfairly on vulnerable groups, such as the young, the elderly or particularly susceptible individuals. Furthermore, exposure to an activity may result in a low level of average risk to any one individual but the totality of such risks across the affected population might not be acceptable as judged by the socio-political response to a particular event such as a railway disaster. Nevertheless, the HSE has suggested that an individual risk of death of 1 in 1000 per annum should *on its own* represent the dividing line between what could be just tolerable for any substantial category of workers for any large part of a working life and what is unacceptable for any but fairly exceptional groups. For members of the public who have a risk imposed on them in the wider interest of society this limit is judged to be an order of magnitude lower – at 1 in 10,000 per annum.

However, these limits rarely bite. Hazards that give rise to such levels of individual risks also give rise to societal concerns and the latter often play a far greater role in deciding whether a risk is unacceptable or not. Second, these limits were derived for the activities most difficult to control and reflect agreements reached at international level. In practice, most industries in the UK achieve at least one order of magnitude better than that.

Risks giving rise to societal concerns

Developing criteria on ToR for hazards giving rise to societal concerns is difficult. Hazards giving rise to such concerns often involve a wide range of events with a range of possible outcomes. The summing or integration of such risks, or their mutual comparison, may call for the attribution of weighting factors for which, at present, no generally agreed values exist (such as the death of a child as opposed to an elderly person, dying from a dreaded cause like cancer or the fear of affecting future generations in an irreversible way).

Nevertheless, the HSE has adopted the criteria below for addressing societal concerns arising when there is a risk of multiple fatalities occurring in one single event. These were developed through the use of so-called FN-curves (obtained by plotting the frequency at which such events might kill N or more people against N). This technique provides a useful means of comparing the impact profiles of man-made accidents with the equivalent profiles for natural disasters with which society has to live. The method is not without its drawbacks, but in the absence of much else it has proved a helpful tool, if used sensibly.[3] Moreover, the criteria are based on an examination of the levels of risk that society was prepared to tolerate from a major accident affecting the population surrounding the industrial installations at Canvey Island on the Thames. These criteria are, however, directly applicable only to risks from major industrial installations and may not be valid for very different types of risk, such as flooding from a burst dam or crushing from crowds in sports stadiums.

Thus, where societal concerns arise because of the risk of multiple fatalities occurring in one event from a single major industrial activity,[4] the HSE proposes the following basic criterion for the limit of tolerability, particularly for accidents where there is some choice as to whether to accept the hazard or not (for example the risk of such an event resulting from a major chemical site or complex continuing to operate next to a housing estate). In such circumstances, the HSE proposes that the risk of an accident causing the death of 50 people or more in a single event should be regarded as intolerable if the frequency is estimated to be more than 1 in 5000 per annum.[5]

Notes

1   See (HSE, 2001a), paragraphs 96–108.
2   HSE enforcement policy for the replacement of iron gas mains available at www.hse.gov.uk/gas/domestic/gasmain.pdf.
3   For a review of the merits and disadvantages of FN curves, see Ball and Floyd (1998).
4   Here a single major industrial activity means an industrial activity from which risk is assessed as a whole, such as all chemical manufacturing and storage units within the control of one company at one location or within a site boundary, a cross-country pipeline or a railway line along which dangerous goods are transported.
5   See HM Treasury (1997) for a discussion of techniques available for extrapolating this criterion to other numbers of casualties and frequencies.

References

Ball, D. J. and Floyd, P. J. (1998) 'Societal risks', report, Risk Assessment Policy Unit, HSE, London

HSC (1991) 'Report of the Advisory Committee on Dangerous Substances: Major hazard aspects of the transport of dangerous substances', HSE, London

HSE (2001a) *Reducing Risks, Protecting People*, HSE Books, Sudbury, UK

HSE (2001b) 'The Health and Safety Executive's enforcement policy for the replacement of iron gas mains', www.hse.gov.uk/gas/domestic/gasmain.pdf

HM Treasury (1997) 'Appraisal and evaluation in central government', *The Green Book*, HMSO, London

House of Commons Select Committee on Defence (2002) *Defence and Security in the UK*, 6th report, July, House of Commons Select Committee on Defence, London

OECD (2001) *Citizens as Partners: Information, Consultation and Public Participation in Policy Making*, OECD Publications, Paris

Office of the Deputy Prime Minister (2003) 'Our fire and rescue service white paper', Cm 5808, HMSO, London

Baroness O'Neill (2002) 'Trust and Transparency', BBC Reith Lecture 4
The Presidential/Congressional Commission on Risk Assessment and Risk Management (1997) 'Framework for environmental health risk management', Final Report, vol 1, The Commission on Risk Assessment and Risk Management, Washington, DC

7

What Makes Tolerability of Risk Work? Exploring the Limitations of its Applicability to Other Risk Fields

Robyn Fairman

Introduction

This chapter explores what makes the concept of 'tolerability of risk' (ToR) work within the British regulatory framework for occupational health and safety. It is argued that for ToR to be successfully operationalized two fundamental components are necessary. The first prerequisite is an acceptance and a legitimization by stakeholders of the need to balance risks being regulated or created against the costs involved in controlling them. The second prerequisite is a form of institutional decision-making that allows that 'balancing of risks and costs' but ensures that decisions are reached. In this chapter I will argue that ToR successfully operates in occupational health and safety in Britain owing to the common law origins of the criminal duties in health and safety law and the corporatist model of social partnership institutionalized in the Health and Safety Commission.

The first part of this chapter explores the acceptance in Britain of the balancing of risks against benefit and examines the historical route from civil cases to 'as low as is reasonably practicable' (ALARP) and the ToR framework. The second part focuses on the institutional setting for decision-making in health and safety. Using the setting of occupational standards for chemicals as an example, the importance of neo-corporatist models of social partnership in the operationalization of ToR will be explored.

ALARP, ToR and Risk Assessment

UK policy-makers have long held the belief that regulation should follow the 'reasonably practical' or 'best practice' rule. The 'best practice' principle was first introduced by the government in 1842 to decrease its involvement in regulation in industry (Ashby and Andersson, 1981; McCormick, 1991). The more modern concept 'as safe as reasonably practicable' was defined in the 1949 court case of Edwards v. the National Coal Board (Asquith, 1949). This case held that the measures that an employer must take to protect worker safety need only be proportionate to the risk the worker was exposed to. This principle of weighing risks and the consequences of accidents and incidents against the costs of preventing them occurring is commonly applied in defining what is 'reasonable' action or a 'reasonable' standard of care in civil tort law. The findings of the courts in legal actions for negligence (the most important tort in health and safety) have created much of the landscape of statute law in Britain. The 1974 Health and Safety Act used this common law principle of 'as low as reasonably practicable' (ALARP) as a way of qualifying the broad criminal duties of employers. In UK legislation and policy, ALARP is the weighing up of the risks with the cost of risk reduction.

ALARP is not a feature of environmental law. In this regulatory arena, the terms 'best practicable means', 'best available technology' and 'best practicable environmental option' are used. Environmental legislation relies on fixed discharge and emission standards, authorizations of polluting processes and nuisance provisions. Food safety legislation is slowly incorporating some risk-based concepts but it is not a central theme of regulation (Codex Alimentarius, 1999). Standards to be achieved in food legislation are more commonly 'safety' or 'wholesomeness'. The principle of ALARP is being put forward as an approach by the UK in European discussions to avoid the implementation of Europe-wide fixed food-safety standards. Risk-based techniques are often used in standard setting, and ALARP will be insinuated into decision-making, but explicit recognition of ALARP is not the central tenet of either environmental or food-safety legislation. This poses a major problem for the incorporation of ToR concepts into these other regulatory arenas. Even environmental nuclear risks which are regulated on the basis of the risk versus cost/feasibility criteria being 'as low as reasonably achievable' (ALARA) are subjected to a 'justification' test before risk concepts are introduced.

The ALARP principle entails, in theory, a simple form of risk–benefit (or cost–benefit) analysis to decide whether the cost of taking a specific action to reduce a risk is justified (HSE, 1988 revised 1992). However, when the Health and Safety Act of 1974 was enacted there were no fixed quantitative guidelines concerning what levels of exposure to risk fitted within the ALARP principles. Following the Sizewell B inquiry, where the inspector, Sir Frank Layfield, recommended a more quantified approach to the ALARP principle as the regulatory agencies did not have any clear risk targets, the HSE published *The Tolerability of Risk from Nuclear Power Stations* as an attempt to attach numerical values to the principle (HSE, 1988 revised 1992).

Following the 1988 HSE report, ALARP risk–benefit analysis now contains a set of basic criteria or exposure limits. Risks that are estimated to be greater than 10^{-4} to the general public and 10^{-3} in the occupational sector should not be considered acceptable and dealt with no matter what the cost (Ball, 1992).[1] Risks that fall between 10^{-4} and 10^{-6} for the public and between 10^{-3} and 10^{-6} for industry should be dealt with on an ALARP basis, and risks that are deemed to be 10^{-6} or less should only be dealt with under particular circumstances (Allen et al, 1992; HSE 1988 revised 1992; Rimington, 1993).

Although originally developed for application to nuclear power, it was envisioned that the ToR framework and numerical criteria would be applied to other health and safety risks (HSE, 1992; Parliamentary Office of Science and Technology, 1996). On the ground on a case-by-case basis, however, the application of these limits is not hard and fast. It is recognized that 'in practice these limits rarely bite' (HM Treasury, 1996). For instance, in the setting of exposure standards for air (either occupational or ambient), no reference is made to these risk criteria. Two crucial factors impede the application of these numerical risk criteria. The first is that much of the agenda-setting for risk regulation in the UK is out of the UK government's control. European legislation will determine priorities, and even where the UK's analysis deems a risk-management action non-cost effective, its hand may be forced, as was the case in the banning of white asbestos, where an imminent European ban meant that the UK also had to ban the substance (HSE, 1999a). The second factor is intimately bound up with the first – some risks are more political than technical in nature, often necessitating regulation beyond the *de minimis* level, no matter what the actual cost may be. There are also the so-called 'popular' risks, often amplified by adverse media exposure. Political imperative here is more important than reference to theoretical risk limits. An example is the enforcement campaign by the Health and Safety Executive concerning faulty gas heaters and carbon monoxide emissions. The numerical risk of death is low (1 in 3.5 million), but the nature of the risk, involving young students living in poor quality rented accommodation, means that the issue is a priority for the enforcement agency (HSE, 1999b).

ALARP is the basis of much UK risk legislation and it characterizes the UK's pragmatic approach of attempting to be practical, flexible and efficient. When compared with the Dutch and German experiences, it is seen to have advantages. The Dutch risk assessment adopts a 10^{-6} highest tolerability level and 10^{-8} *de minimis*, with everything in between on the ALARA principle. Several questions on these guidelines have been raised by UK risk experts. Many feel that the legislation is only suitable for certain situations and is only likely to be useful for public relations purposes (O'Riordan, 1985; Wynne, 1992). The cost of adopting 10^{-6} across-the-board as the lowest possible tolerable level is likely to be high and it offers little opportunity for flexibility (Ball, 1992; Layfield, 1987), and the costs of enforcing across-the-board 10^{-6} are likely to be prohibitive (Ball, 1992; Layfield, 1987).

ALARP also has problems, however. Wynne (1992) states that there are many uncertainties in using ALARP. The lack of inclusion of social and environmental externalities in the ALARP calculations is one problem. This is difficult, however, as economic instruments used to measure externalities have, to date, completely failed (Stirling, 1992 and 1997). What is ALARP to those creating the risk (and consequently gaining benefit from it) is often different from ALARP as defined by workers (who also gain some benefit), which is again different to ALARP as defined by those who only experience disbenefit from the risk. Interpretation and application of ALARP at the European level has also been problematic. The British Trade Unions lobbied the European Commission to take action against the UK for the way that the health and safety framework and daughter directives on occupational health and safety were being translated into UK legislation and the inclusion of ALARP as a qualifier to obligations. Application of ALARP has also been questioned on a sector basis (Woolfson and Beck, 1995) and within individual firms (Dawson et al, 1988).

The development of the ToR framework arose out of the consideration of ALARP. ALARP has also meant that risk is central to health and safety policy and regulation in Britain. The HSC and HSE have a long history of risk-based approaches and risk assessment. Risk concepts are central to British health and safety legislation. The term is used in its lay meaning in the general duties of the 1974 Health and Safety at Work Act (HSWA). This lay interpretation was confirmed by a Court of Appeal case in which the meaning of risk in the HSWA was interpreted as meaning 'the possibility of danger' (R v. Board of Trustees of the Science Museum, 1993, 1 WLR 1171). The 1999 Management of Health and Safety at Work Regulations require all businesses to carry out risk assessments of their activities.

Risk is a fundamental concept in the regulation of occupational health and is incorporated into legislation. This is not common across other regulatory fields or sectors but without it ToR concepts cannot be applied. For ToR to operate, stakeholders must be happy with the concept of risk as a measure used in regulation. Many consumer stakeholders demand safety and are not happy with the concept that risk should be weighed against cost considerations. In food regulation, for instance, safety is the objective, and the consumer panel of the Food Standards Agency regularly questions whether cost is a valid topic for consideration (Consumer Committee, 2004). Legislative schemes in food safety do not tend to allow for the consideration of risk and the weighing of risk against costs. For instance, residues of pesticides in food are not allowed if any risk exists – if residue concentrations are higher than the threshold for health effects, either the product is banned or Good Agricultural Practice Controls altered to ensure that safety levels are not breached.

For ToR to operate there needs to be an explicit acceptance of the importance not only of risk, but of the costs incurred in reducing risk. This acceptance is not common in other regulatory frameworks. In environmental protection, balancing concepts are used, but these balance the technology available with the costs of implementation (BAT or BATNEEC) rather than risk. In newer environmental regulation – for instance with regard to contaminated land – risk-based standards are being used and cost is considered. However this is not as a direct balance against risk, and minimum standards must be achieved. There is scope for forms of ToR to be operated in these new legislative regimes if the second pre-requisite for its operation can be tackled – that of how you get stakeholders to 'buy in' to a balancing of interests.

Enabling Safety Policy Decisions – Corporatism in Action

In the recent literature of political science and sociology, the term 'neo-corporatism' refers to social arrangements dominated by tripartite bargaining between unions, the private sector (capital) and government. Most political economists believe that neo-corporatist arrangements are only possible in societies in which labour is highly organized and various labour unions are hierarchically organized in a single labour federation. Such 'encompassing' unions bargain on behalf of all workers.

A major factor in why ToR works is the corporatist policy-making institutions for health and safety established by the HSWA. Corporatism has at its core the advocacy of an institutional relationship between systems of authoritative decision-making and interest representation (Molina and Rhodes, 2002). This process was a feature of a Keynesian framework in which:

> *the major interest groups are bought together and encouraged to conclude a series of bargains about their future behaviour, which will have the effect of moving economic events along the desired path. The plan indicates the general direction in which the interest groups, including the state in its various guises, have agreed that they want to go.* (Shonfeld, 1965)

A crucial distinction between pluralism and neo-corporatism is the extent to which organized groups are integrated into the policy-making arena of the state (Martin, 1983). Mansbridge (1992) argues that corporatist theories give a richer account of deliberation within interest groups than do pluralist theories. Corporatism postulates a well-regulated framework of interaction where neither the state nor the interest groups lose sight of public interest considerations. In a pluralist environment populated by competitive, self-seeking interest groups, there is little room for balancing interests (Hunold, 2001).

ToR works because it is a principle made operational within a policy-making system that:

- limits the number of actors in the discussion;
- binds the actors involved to sell the position to their constituents; and
- forces the balancing of interests.

An important factor is that, unlike many other areas of public policy dispute, all the interest groups involved in health and safety have an overriding common objective. The long-term stability and wealth of industry and its activities is fundamental to all interest groups, be they industry (increased shareholder capital), the state (economic growth and stability) or trade unions (job security and wage stability).

These issues will now be discussed by looking at a case study of how health and safety policy-making occurs in practice. To assist in the discussion of how this may relate to the operation of ToR, an institutional analysis is performed examining the 'rules of the game' (Ostrom, 1986). This helps develop the argument that corporatist institutional structures are vital to the operation of ToR.

Case Study: Setting Standards for Chemicals in Britain

The institutional setting: The UK Health and Safety Commission Executive

The HSC is a tripartite organization run by ten commissioners appointed by the Department of Work and Pensions and representing a set of organizations that includes local authorities, employees and employers. It is, in effect, run by stakeholders in health and safety. The Chair is currently a trade union representative. The HSC is an example of corporatist style of decision and policy-making. The HSC was established by the HSWA, the product of the Robens Committee of Inquiry (discussed in Chapter 4). This inquiry and legislation are products of the style of industrial relations operational in the late 1960s and 70s. This was the time of the revival of the theory of the corporate state. Developments in all Western societies suggested that public decision-making was increasingly becoming a tripartite affair of bargaining between the state, employers' associations and trade unions. Corporate bodies representing functional interests were being incorporated into the machinery of the state, complementing and to some extent replacing formally representative bodies such as Parliament. In return for a share in the making of political decisions, the non-state organizations were expected to be able to discipline their members and ensure that they would support the agreed policies. The HSC and HSE are among the functioning embodiments of institutional corporatism.

Although other ventures into the social contract with the trade unions have failed (mainly owing to the inability of the Trades Union Congress (TUC) to control its member unions), the HSC is seen as successful from all participants' points of view.

The HSC's primary function is the safety and welfare of employees and the public affected by work activities in the UK. It conducts research, proposes new laws and standards, and provides information. The HSC receives assistance and advice from its operational arm the HSE, which has a staff of 4500, ranging from scientific and medical experts to technicians, policy advisers and inspectors. The broad focus of the HSE is to give advice and information on regulatory policies and to enforce regulation (when necessary) in industry that is otherwise self-regulating.

In setting chemical standards for worker safety, the HSC relies on the advice of expert advisory committees and working groups.

The Advisory Committee on Toxic Substances (ACTS)

ACTS is a tripartite committee whose remit is to advise the HSC on matters relating to the prevention, control and management of hazards and risks to the health and safety of persons arising from the supply or use of toxic substances at work. ACTS is also concerned with the risks posed to consumers, the public and the environment. It is a general risk-management committee and has local government and consumer members. The corporatist style extends into the advisory committees of the HSC. The HSC advisory and scientific committees are fundamentally different from advisory and scientific committees in the rest of UK government as they are tripartite expert committees built on a corporatist model. Stakeholders such as the employees and employers' representatives nominate experts to the HSC advisory committees who argue from their interest group position. In relation to worker standards for chemicals, ACTS considers the recommendations of the expert committee WATCH (see below). There are two types of recommendations – either the chemical will be given a health-based standard (where it is feasible and cost-effective to do so) or they will set a standard that weighs risk against cost and feasibility (for instance for carcinogens or chemicals it is unfeasible or very costly to reduce to a 'safe' level).

Working Group on the Assessment of Toxic Chemicals (WATCH)

WATCH is a tripartite sub-committee of ACTS. WATCH comprises toxicologists/epidemiologists and occupational hygienists (four TUC scientists, four Confederation of British Industry (CBI) scientists, five or six independent scientists and five or six HSE toxicologists/occupational hygienists). It considers the evidence on the occupational exposure and health effects of substances and recommends standards to ACTS. It gives views to the HSE on its contribution to occupational exposure, hazard and occupational risk assessments in its role as a partner in the UK Competent Authority for the EU Existing Substances Regulation. WATCH recommends to ACTS measures that seem to be appropriate in the light of these assessments. WATCH reviews scientific aspects of hazard assessments of substances and other matters referred to it by the HSE, ACTS or the Standing Committee on Hazard Information and Packaging.

ACTS accepts that WATCH has properly assessed the toxicological data and does not re-examine. The TUC and CBI members on WATCH report the activities of WATCH to their ACTS members to give them insight into the basis of the assessment and any problem issues.

The 'rules of the game' in public risk management

The underlying rules by which risk management regimes operate can also be analysed. The Royal Society developed a six-dimensional model from a framework originally developed by Ostrom (Ostrom, 1986; The Royal Society, 1992). The six rules encapsulated the essential 'rules of the game' involved in institutionalized decision-making and are described in Table 7.1, along with the overall direction of change in risk management in relation to them.

Boundary rules are rules defining who has access to the risk management process and who can be counted as a player. *Scope rules* govern what comes within the province of risk management and the limits of what risk-managing institutions may decide. Often the scope of risk-managing institutions is debated. *Position rules* identify the decision points in the risk management process (for example who decides whether to take a prosecution and on what charges), how decision points are arrayed in terms of hierarchy or precedence (local authorities against the HSE, for example), and how individuals are appointed or dismissed from these positions. *Information rules* specify who is entitled to what information from whom and under what conditions, for example in rules of confidentiality and secrecy in requirements to inform the public or employees of risks. *Authority and procedural rules* define how decisions must be made in specifying the order or timing of decisions and what constitutes 'evidence'. Authority and procedural rules may determine where the emphasis is laid between quantification of risk and qualitative arguments. Such rules can determine whether decision advice procedures are to be adversarial or inquisitorial. They may also determine whether decision advice procedures are formal, with elaborate constraints on the way information is presented, or informal, with much fewer constraints on admissible information. *Preference-merging rules* define the ways in which the individual players or stakeholders concerned in the assessment and management of a particular risk are to come together into a collective decision. Preference-merging procedures can be generally distinguished according to whether they have the integrative effect of pulling a community together by arriving at a

Table 7.1 *Dimensions of risk management (after Ostrom, 1986) and putative trends*

Rule Type	Explication	Range of Key Types	Characteristic Trends
Boundary	Who is counted as a player?	Technocratic/participative	More participative
Scope	What is managed and what can be decided?	Broad/narrow	Extension of scope
Position	What is the hierarchy of the players?	Single organization/ multi-organization	More multi-organizations
Information	Who is entitled to know what from whom?	Open/closed	More open
Authority and procedure	Under what conditions must decision be made?	Formal/informal	More formal
Preference merging	How are individual preferences aggregated into collective decisions?	Consensus (integration)/ conflict (aggregation)	More conflict

Source. Royal Society (1992), p149

consensus or an aggregate effect of dividing a community by procedures that give the decision to one party at the expense of another.

Organizational rules in setting chemical standards within the HSC and HSE

Boundary rules

In occupational safety at both UK and EU levels the major stakeholders have specific roles. The Advisory Committee for Safety, Hygiene and Health Protection at Work (ACSHH) and the HSC and its advisory committees are tripartite. The 'peak' stakeholders in safety manage the HSC and provide its direction and planning. Although access to the expert committees is restrictive and a member of the public could not attend and participate, the ultimate in participative boundary rules, the process is participatory in that stakeholders nominate the participants in the decision-making bodies.

Scope rules

The HSC and HSE have a wide scope in managing the risks arising from work. These could be, for example, risks to workers, risks to the public or risks to subcontractors. The organization covers all types of hazard arising from work, ranging from radiation, biological and chemical risks to psychosocial risks. This ensures that the authority and practice of the stakeholders is not challenged by others with competing philosophies.

Position rules

The EU is the primary legislator in occupational safety and health terms. The setting of indicative values by the European Commission allows the HSC to set its own standards with regard to European levels. WATCH and ACTS use this flexibility. WATCH reviewed the EU indicative standard on diethyl ether, for example. The EU's recommendation was 100ppm, but WATCH made a decision of 400ppm because it was not felt reasonably practicable for UK industry to meet the tighter standard.

Authority rules

Although the framework for standard setting is very formalized (through legislation and European monographs), no standardized procedures exist for dealing with data. No formal methods are used; models and uncertainty factors are based upon specific circumstances of each case. This allows committees to operate mutually agreed procedures on a case-by-case basis.

Informational rules

All minutes of ACTS and WATCH meetings are published on the internet. Once a proposal is made, there is wide consultation on the proposed standard, with details of the factors, evidence and analysis that is behind the standard published and open to consultation. The members of committees are nominees of stakeholders, so the process is effectively open, with open lines of communication. Making the process more apparently open though information is interesting in light of the corporatist structure, since other stakeholders have no way to influence the process directly.

Preference merging

The objective of ACTS and WATCH is to arrive at a consensus decision. Although all are appointed:

> *in most discussions it is hard to say that anyone has an agenda. In some discussions a bias is evident. In these instances the consensus decision is arrived at, with the minority views clearly recorded.*
> (WATCH Member)

An interesting point about the consensus objective was made in the TUC's response to the HSE consultative document *Reducing Risk, Protecting People* (HSE, 1999b). It states that it agrees that the approach in the consultative document describes the process used by the HSC and HSE to determine the appropriate systems for the control of workplace risks but argues that:

> *the approach is the product of the consensus on which the HSC is based. Unions and employers, to take just two stakeholders, make*

decisions about the control regimes we find appropriate on different bases to those described in the document.

Our perception is that employers will seek a risk control regime that allows them to maintain an acceptable system of work at the lowest possible cost. [...] But it produces a theoretically quite different approach to that of trade unions, where the question of cost is far less central. [...]

These different perspectives mean that the employers' and employees' representatives have two quite different patterns of discourse on safety. Employers want to reach a position which is 'safe enough', while unions will always want to move beyond that to a position which is merely 'safer' than the current position.

These tensions do not mean that employers and unions can never agree [...] and the process of the HSC produces a consensus which does deliver both, albeit momentarily. That consensus is what is described in Reducing Risks, Protecting People, *but because it does not recognize the dynamic tension between the stakeholders, the discussion document is only really good at describing how a particular control regime is arrived at.* (Tudor, 1999)

A 'dynamic tension' exists in the HSC and its advisory committees. The outcomes of the HSC are consensus decisions but they are arrived at through a dynamic process. The members of ACTS and WATCH know each other personally and recognize the expertise and experience of each member. In the opinion of a member of the WATCH committee:

After a while you see that arguments are based on people's experience rather than the position they occupy. (WATCH 1999 Member)

When data is lacking, arguments about how to interpret the situation tend to be more problematic. The judgement of the group fills in the data-gaps:

In meetings you get arguments, marked disagreements. These get talked through until you get a genuine understanding of the issues. Sometimes confrontation is based on individuals' interpretation of data. Sometimes there may be other reasons and issues, especially when the data is lacking. General experience of the group helps to fill in these gaps. (WATCH 1999 Member)

A member of ACTS commented that in the past ACTS could be very confrontational, but that the committee is now much more professional. Although consensus is the aim, the process of obtaining it can be adversarial.

ToR and its use in setting chemical standards

The HSE uses the Tolerability of Risk (ToR) framework as a basis of the criteria for deciding whether risks are unacceptable, tolerable or broadly acceptable. The ToR framework is based on ethical, utility and technology criteria (HSE, 1999b). According to the HSE the strength of the framework lies in its combination of all three types of criteria and the fact that the framework is very similar to the approach that people apply in everyday life:

> *In everyday life there are some risks that people choose to ignore and others they are not willing to entertain. But there are also many risks that people are prepared to take by operating a trade-off between the benefits of taking the risks and the precautions we all have to take to mitigate their undesirable effects.* (HSE, 1999b)

Risk encompasses more than physical harm, taking into account other factors such as ethical and social considerations. The ToR framework can be seen as applying ethical criteria at the highest levels of risk, where the risks are unacceptable. In the middle, tolerable region, utility-based criteria dominate. In this region, the HSE believe that people are prepared to tolerate the risks in order to secure benefits (such as employment or lower costs of production) in the expectation that:

- *the nature and level of the risks are properly assessed and the results made available. The assessment of risks needs to be based upon the best available evidence and, where evidence is lacking, the best available scientific advice;*
- *the risks are not unduly high and should be kept as low as is reasonably practicable (the cost of risk reduction will be compared to the risk); and*
- *the risks are periodically reviewed to ensure that they continue to meet the ALARP criteria.* (HSE, 1999b)

The ToR framework forms the basis of tolerability decisions in setting standards. Occupational exposure standards (OELs) fit within this framework. The health-based standard should ensure that the risks are within the 'broadly acceptable' region. The standards determined by risk versus cost and feasibility are legal limits and as such they must lie on the boundary of barely tolerable and unacceptable. This exposes, however, the inconsistency and difficulties in applying numerical criteria to the ToR framework. The numerical levels of risk that apply to these regions of unacceptability, tolerability and acceptability are likely to be very different for different hazards. For instance, in numerical terms the upper limit of tolerability for the death of a worker is 1 in 1000 per annum (HSE, 1992). The risk of death from being exposed to above the maximum exposure level (MEL) for hardwood dust is estimated to be much lower than this. The tolerability limits exclude other issues that are important in risk tolerability.

Institutional factors in the success of ToR in occupational health and safety

The case study of how standards are set shows that stakeholder involvement is institutionally built into the setting of chemical standards. In the tripartite nature of the HSC, ACTS and WATCH, the main stakeholders not only have a say in the process, but also are responsible for the setting of standards. This is at the heart of corporatist models of social partnership. The system is accepted and praised by both employers' and employees' representatives. In its response to the review of the May Principles (May, 1997) on the use of scientific advice in policy-making, the TUC stated:

> *We have often recommended the system adopted by the HSC in obtaining scientific advice on the regulation of toxic chemicals – a panel of scientific experts has been established, with the main stakeholders (unions, employers and government) all making nominations. Backed up by the HSC's extensive public consultation process, this process ensures that the potential bias* [of scientific advice based on partial evidence] *can be identified and considered critically.* (Tudor, 2000)

The CBI has also praised the HSC, making a case that environmental regulation should be established on the same principles (CBI, 1998). Participation wider than these three stakeholders has also been attempted. WATCH at one time had an environmental/consumer member, but this person had real difficulty in being able to take part in meetings owing to their technical nature.

If we consider how government makes decision about risk in other fields, the use of scientific expert groups is common. Technical expert bodies aim for consensus decisions. The government needs a single conclusive recommendation on which to base decisions. This is necessary so that government can be seen to be acting on sound scientific advice. Scientific committees view the issues in terms of their training and expertise and confine the issues to those within their purview. Much of the work of the expert groups in risk issues is the 'post normal' science as described by Funtowicz and Ravetz (1996). The science is uncertain, studies are of variable quality, values are disputed, stakes are high and decisions uncertain. In such circumstances, Funtowicz and Ravetz recommend an extended peer review and active involvement of stakeholders and lay persons.

The changes to advisory committees (DH, 1999) and reviews on the operation of scientific committees (UK FSA, 2000) are a step towards this active involvement, although it is questionable what the impact a single layperson will have in a technical committee. A question must be raised as to how effective the use of expert committees with a limited number of stakeholders or lay persons is as a participative process. One or two voices, representing bodies with differing power bases, among a large number of experts are unlikely to result in a consensus decision. It is more likely that the decision will be a compromise on behalf on some constituent group (Salter, 1988). However, experiences in the inclusion of lay people in the Committee on Toxicity and the Advisory Committee on Releases to Environment show that lay persons do change the process of committees (Löfstedt and Fairman, 2006). The real question is whether this makes for better decisions. UK government departments are currently taking on the arguments about participation and open government. Wider participation by stakeholders in risk assessment and management may expose hidden assumptions and errors, but would the resulting discussion lead to a decision? The result of attempting to reach a decision where stakeholders hold different (and often competing) interests may be no consensus. It could be said that this would air value-differences and allow discourse. Alternatively it could result in stakeholders becoming involved in an adversarial fight and further entrenching their various positions (Pidgeon, 1996). As Jasanoff states in her comparison of UK and US cancer policy:

> *Comparing consensus-building mechanisms across several countries shows that people holding widely divergent scientific or political views can be brought together only with great difficulty, if at all. Scepticism about the meaning of 'consensus' under these circumstances is thoroughly justified. Yet in spite of their imperfections and limitations, mechanisms that encourage negotiation offer the greatest promise of allowing participation without bringing government to a halt.* (Jasanoff, 1986)

ToR may be held up by some as a way of overcoming risk-management problems and enabling more 'rational' risk-management decisions. However, ToR works in the HSC and HSE because of its corporatist structure, not because of any amazing insight on behalf of the ToR framework. The HSC committees, built on tripartite principles, address the issue discussed above of 'post normal' science head on. In these committees the different stakeholders in the process appoint members. Issues of values and the judgements and assumptions that will flow from these are acknowledged. Although the process aims for consensus, the comments by the TUC on the active adversarial nature of the meetings show that value issues are important.

So if corporatism works in health and safety policy, could it (and ToR) not be applied to other risk areas? In occupational health and safety issues, although traditionally seen as being in opposition over many issues, both employers' and employees' representatives have a common goal: they both want to keep companies in business. From the employers' view they wish to maintain profits; from the employees' view they wish to maintain jobs. In politically controversial areas such as food safety or environmental protection, there is often no common goal. In food safety, for instance, consumers want safe, cheap food. Food producers want maximum profits. The risks of food production (for example pesticides or genetic modification) are borne by the consumer or environment, the profit made by the food producer. It is difficult to see how a technical expert group with members appointed by the food industry, consumer groups and government would overcome basic value issues on the use of pesticides, for instance, to be able to scientifically assess safety. The same problems apply to environmental protection. The producers of pollution benefit from manufacturing while those affected by pollution suffer the risk. It is hard to see how consensus could be achieved through such a process. Even within the occupational health and safety arena, on some issues and in some areas consensus is achieved, while statements have to be made defining the different positions in others.

The application of neo-corporatist theory outside industrial relations or economic policy areas and interest organizations representing capital, labour and the state is rare. The Australian government used a neo-corporatist approach to interest-group intermediation when developing a sustainable development process (Downes, 1996). It included peak interest organizations involved in environmentalism and development issues in an exclusive consultation process. The outcomes were not successful, however, owing to the difficulties in ensuring a single voice from organizations and strong enough incentives to cooperate.

McFarland, who examined the political behaviour of public interest groups, recounts another example of neo-corporatism in environmental issues (McFarland, 1993). He discussed the National Coal Policy Project in the US – a two-year effort on the part of 60 representatives of environmental organizations and business firms to reach agreement on federal regulations governing the production, transportation and burning of coal. Funded by both foundations and companies, its purpose was to see if representatives of two often hostile sets of interest groups could agree on a common set of policies that would further both of their respective objectives, namely the preservation and improvement of environmental quality *and* the development of coal as an energy source.

On one level, this experiment in business–environmental cooperation was extremely successful. The project's participants came to better understand, learn from and respect each other and were able to reach agreement on a wide range of issues. From another perspective, however, this experiment in 'cooperative pluralism' or regulatory corporatism was a failure because neither side could deliver their respective constituencies. A number of important environmental organizations, including the Sierra Club and the Natural Resources Defense Council, refused to allow representatives of their organizations to participate in the project. Their refusal stemmed from a variety of factors: they mistrusted the motives of industry in wanting to work with them, they did not wish to limit their political options and they reasoned they could get a better deal by taking on the industry rather than negotiating with it. Nor were a critical mass of individual firms or trade associations willing to be bound by the project's recommendations, let alone invest their scarce political resources in lobbying for their adoption.

This does not mean that cooperation between business firms and environmental organizations is not possible, but it does suggest that cooperation is only possible when its objectives are modest and clearly in the interests of each organization. In many cases, this cooperation is as likely to undermine as to advance the public interest. This is probably the fundamental point of why corporatism in UK health and safety regulation allows decisions to be made and trade-offs between risk and cost agreed. The objectives of all sides are similar at heart, and all sides win by being part of the process of decision-making. The TUC and CBI will actively work at 'delivering' their constituents. It is hard to see what common objectives might exist between the stakeholders in food safety or environmental protection, nor whether the benefits of being allowed to sit at the table would be enough to force the peak stakeholders into working at delivering their constituents. The interest groups are diverse and multitudinous in these risk areas and peak interest groups do not often exist. Even if they did, the diversity of groups makes the development of common positions hard, if not impossible. The case of occupational health shows that only three stakeholders are invited to participate. If they moved to a more pluralist policy-making, there are many interest groups in the field that would actively oppose ToR concepts (for instance the Campaign for Corporate Accountability, the Construction Safety Campaign and several trade unions).

Conclusion

ToR is a framework that has a long history and is accepted by employers' and employees' representatives. The concept is at the heart of the regulation of safety in the UK and derives from common law interpretations of what is 'reasonable'. It is operated within a corporatist model of social partnership for health and safety where the stakeholders are expected to deliver their constituents in the implementation of the objectives of the organization. Risk within the HSC and HSE is recognized to be a mixture of science, values and judgements, and the tripartite structures of decision-making enable this concept of risk to be used. The scientists on expert groups such as WATCH can examine risk from their value positions (they are appointed by constituent stakeholders). They make judgements on risk assessment aware of the value issues involved. The importance of this cannot be overstated: scientific judgement does not become a battlefield – the scientists come to a consensus decision and present that to their stakeholder sponsors. Arguments about science and risk occur within the committee, but the decision will not then be challenged by stakeholders or interest groups as 'their' scientists made them. ToR is vital in occupational safety and may have something to offer other policy areas. However the stakeholders in the risk arena under question must accept risk as a central concept (not common in consumers), must be willing to balance 'safety' against costs (also not common in consumers) and must be tied into a policy-making regime where they are forced to accept compromises. Attractive as the concept is to industrial stakeholders, it will be less acceptable to consumers.

Note

1   For those unfamiliar with the notation: a risk of 10^{-3} means 1 in 1000, 10^{-4} means 1 in 10,000, and so on.

References

Allen, F., Garlick, A., Hayns, M., and Taig, A. (1992) *The Management of Risk to Society from Potential Accidents*, Elsevier Applied Science, London
Ashby, E. and Anderson, M. (1981) *The Politics of Clean Air*, Clarendon Press, Oxford
Asquith, Lord (1949) in Edwards *v.* National Coal Board (1949) 1 KB; 1949, 1 All ER743, 712 and 7477, a case of interpretation of S 102 (8) of the Coal Mines Act 1911
Ball, D. (1992) 'Understanding the risks', *Chemistry and Industry*, no 20, pp776–779
Codex Alimentarius Commission (1999) *Conference on International Trade Beyond 2000: Science Based Decisions, Harmonization, Equivalence, and Mutual Recognition*, Alicom9: report of the 25th conference
Confederation of British Industry (1998) *Worth the Risk: Improving Environmental Regulation*, CBI, London
Consumer Committee (2004) *Protecting consumers: The FSA and Regulation*, Paper to FSA Consumer Committee, 16 December
Dawson, S., Willman, P., Bamford, M. and Clinton, A. (1988) *Safety at Work – The Limits of Self-Regulation*, Cambridge University Press, Cambridge
Downes, D. (1996) 'Neo-corporatism and environmental policy', *Australian Journal of Political Science*, vol 31, no 2, pp175–190
Department of Health. (1999) *Improved Arrangements for Openness of Committee on Toxicity, Committee on Mutagenicity, and Committee on Carcinogenicity of Chemicals in Food, Consumer Products and the Environment*, Department of Health, London
Funtowicz, O. and Ravetz, J. (1996) 'Risk management, post-normal science and extended peer communities', in C. Hood and D. Jones (eds) *Accident and Design*, UCL Press, London
HSE (1988, revised 1992) *The Tolerability of Nuclear Power Stations*, HMSO, London
HSE (1999a) *Consultation on the Prohibition of Asbestos Regulations*, HSE Books, Sudbury, UK
HSE (1999b) *Reducing Risks, Protecting People*, HSE Books, Sudbury, UK
Her Majesty's Treasury (1996) *The Setting of Safety Standards*, HM Treasury, London
House of Lords Select Committee on Science and Technology (2000) *Science and Society*, HMSO, London
Hunold, C. (2001) 'Corporatism, pluralism and democracy: Towards a deliberative theory of bureaucratic accountability', *Governance: An international journal of policy and administration,* vol 14, no 2, pp151–167

Jasanoff, S. (1986) *Risk Management and Political Culture*, Russell Sage Foundation, New York

Layfield, Sir Frank (1987) *Sizewell B Public Inquiry Report*, HMSO, London

Löfstedt, R. and Fairman, R. (2006) 'Scientific peer review to inform regulatory decision-making: A European perspective', *Risk Analysis*, vol 26, no 1, pp25–31

Mansbridge, J. (1992) 'A deliberative theory of interest representation', in G. E. Marcus and R. L. Hanson, *Reconsidering the Democratic Public*, The Pennsylvania State University Press, University Park, PA

Martin, R. (1983) 'Pluralism and the new corporatism', *Polit Stud.*, no 31, pp86–102

May, R. (1997) *The Use of Scientific Advice in Policy-Making*, Parliamentary Office of Science and Technology, London

McCormick, J. (1991) *British Politics and the Environment*, Earthscan, London

McFarland, A. (1993) *Cooperative Pluralism: The National Coal Policy Experiment*, University Press of Kansas, Lawrence, KS

Molina, O. and Rhodes, M. (2002) 'Corporatism: The past, present and future of the concept', *Annual Review of Political Science*, vol 5, pp305–331

Ostrom, E. (1986) 'A method of institutional analysis', in F. Kaufman, G. Majone and V. Ostrom (eds) *Guidance, control and evaluation in the public sector*, Walter de Gruyter, Berlin

O'Riordan, T. (1985) 'Approaches to regulation', in H. Otway, and M. Peltu (eds) *Regulating Industrial Risks*, Butterworths, London

Pidgeon, N. (1996) 'Technocracy, democracy, secrecy and error' in C. Hood and D. Jones (eds) *Accident and Design*, UCL Press, London

Parliamentary Office of Science and Technology (1996) *Safety in Numbers – Risk Assessment in Environmental Protection*, POST, London

Rimington, J. (1993) *Coping with Technological Risk: A 21st Century Problem*, The Royal Academy of Engineering, London

Royal Society (1992) *Risk: Analysis, Perception and Management*, Royal Society, London

Salter, L. (1988) *Mandated Science*, Kluwer, Dordrecht, The Netherlands

Shonfeld, A. (1965) *Modern Capitalism: The Changing Balance of Public and Private Power*, Oxford University Press, Oxford

Stirling, A. (1992) 'Regulating the electricity supply industry by valuing environmental effects', *Futures*, vol 24, no 10, pp1024–1047

Stirling, A. (1997) 'Limits to the value of external costs', *Energy policy*, vol 25, no 5, pp517–540

Tudor, O. (1999) 'Letter to HSE on the TUC response to reducing risk', *Protecting People*, 22 November

Tudor, O. (2000) 'Letter to Office of Science and Technology on review of guidelines on the use of scientific advice in policy-making', publicly available comment letter, 17 January

UK Food Standards Agency (2000) *Report on the Review of the Scientific Committees*, FSA, London

Woolfson, C. and Beck, M. (1995) *The Deregulation of the British Continental Shelf*, Offshore Information Centre, Aberdeen, UK

Wynne, B. (1992) 'Public understanding of science: New horizons of halls of mirrors?', *Public Understanding of Science*, vol 1, pp37–43

Conclusion

Frédéric Bouder, David Slavin and Ragnar Löfstedt

At the end of this theoretical and practical journey into the concept of tolerability of risk (ToR), a number of key elements have been uncovered. First of all, Renn sets out that the concept of ToR is not only a way to make decisions about acceptable risks, but that it should be understood in the context of theoretical debates about risk governance. In that sense, although it has been primarily defined and applied by practitioners from the HSE, it is far more than a practitioners' tool. It is an integral part of the wider and more sophisticated architecture of risk governance. The contributions from McQuaid and Bandle, on the other hand, illustrate that ToR is linked to specific historical and policy developments that took place within the UK under specific circumstances: it was originally invented to handle risks from nuclear power generation. The success of the concept, which has since been applied to a variety of health and safety situations, shows that the ToR concept is undoubtedly able to deliver.

But beyond the UK experience, how far can ToR inspire regulators?

First of all a ToR-like model would need to be tested against a number of empirical challenges, especially the need to formulate reliable mechanisms to keep constant track of the evolving nature of the perception of risk. Although quantitative estimates about the probability of the occurrence of a particular hazard are relatively stable or even subject to improvement (as a result of technology improvement), public perception about risk is much more fluctuating. In many instances there is a tendency towards lesser degrees of risk acceptance among substantial parts of the public, as their environment is becoming safer. This suggests that a framework like ToR, which takes on board both statistical estimates about the occurrence of hazards and societal concern about risk, is inherently instable, or at least fluctuating. In most cases the application of a ToR-like framework will require regular attention to be paid to the evolution of 'tolerability lines', which could generate substantial costs and would require the designing of a sophisticated tool to appreciate risk perceptions. And perceptions about risk may be less easy to quantify than the technical probability of hazards.

For Fairman, a straightforward limitation is related to the Europeanization of UK law. The growing importance of EU law as a source of regulation implies that non-EU concepts will have to pass a test before they could be introduced. More fundamentally, it has been pointed out that ToR is by essence undermined in a number of 'highly conflictual areas' like food safety or environmental protection. Similarly, Bandle notes, the concept struggles when risks are pervasive and difficult to characterize. The use of the ToR framework has been particularly successful in relation to acute risks, such as accidents, particularly those arising in nuclear and on-shore and off-shore major hazards. With such acute risks, the harmful consequence of the risk is easy to define and estimate – it is usually death. But all risks do not fall into this category. There are, for example, health risks where the harm is more pervasive, there are situations where the effects of an adverse event are delayed and involve chronic diseases, for example in the case of asbestos contamination.

Through revising the concept more systematically, Le Guen and, finally, Fairman provide some insights into the potential use of a ToR approach beyond its current scope. Le Guen stresses the pragmatic value of the concept, which allows putting regulatory transparency and stakeholders' involvement at the core of risk-decision processes. This proves to be particularly important in situations where more participation is needed. Fairman, on the other hand, focuses on the prerequisites that any framework that might try to emulate ToR would probably need to meet:

- an acceptance and a legitimation by stakeholders of the need to balance risks being regulated against the costs involved in controlling that risk; and
- a form of institutional decision-making that allows for the balancing of risk and costs but ensures that decisions are reached.

The potential scope of further application of ToR will remain a matter of debate, and the value of the contribution that we have presented in this volume is its contribution to opening this debate. Outside the UK relatively similar concepts have already been developed, especially in the Netherlands (Ale, 2005) and Scandinavia (Okstad and Hokstad, 2001). Moreover, ToR seems also to have inspired industry-led voluntary approaches, in particular initiatives stemming from insurance industries. It could also be a source of inspiration in the medical field. The formulation of ToR-like models will imply looking carefully at institutional and technical obstacles, but, based on the UK experience, this does not seem to be out of reach. Identifiable reasons for the success of ToR are that it limits the number of actors in the discussion to a manageable number, that it is not blind to public perceptions and that it binds the actors involved to sell a workable position to their constituents. At an institutional level, we would therefore stress the importance of a policy-making context that is sufficiently open to pay attention to public views while managing and containing disagreements among stakeholders in order to reach consensus. This, we believe, represents a considerable set of assets for what could easily be mistaken for narrow technical guidance.

References

Ale, B. J. M. (2005) 'Tolerable or acceptable: A comparison of risk regulation in the United Kingdom and in the Netherlands', *Risk Analysis*, vol 25, no 2

Okstad, E. and Hokstad, P. (2001) 'Risk assessment and use of risk acceptance criteria for the regulation of dangerous substances', proceedings of the European Conference on Safety and Reliability – ESREL 2001, Politecnico di Torino, Italy, 16–20 September, vol 1, pp117–124

Index

Printed and bound by CPI Group (UK) Ltd, Croydon, CR0 4YY

23/10/2024

01777675-0016